Historical & Cultural Astronomy

The Historical & Cultural Astronomy series includes high-level monographs and edited volumes covering a broad range of subjects in the history of astronomy, including interdisciplinary contributions from historians, sociologists, horologists, archaeologists, and other humanities fields. The authors are distinguished specialists in their fields of expertise. Each title is carefully supervised and aims to provide an in-depth understanding by offering detailed research. Rather than focusing on the scientific findings alone, these volumes explain the context of astronomical and space science progress from the pre-modern world to the future. The interdisciplinary Historical & Cultural Astronomy series offers a home for books addressing astronomical progress from a humanities perspective, encompassing the influence of religion, politics, social movements, and more on the growth of astronomical knowledge over the centuries.

More information about this series at http://www.springer.com/series/15156

Wayne Orchiston • Peter Robertson
Woodruff T. Sullivan III

Golden Years of Australian Radio Astronomy

An Illustrated History

 Springer

Wayne Orchiston
Centre for Astrophysics
University of Southern Queensland
Toowoomba, QLD, Australia

and

National Astronomical Research
Institute of Thailand
Mae Taeng, Chiang Mai, Thailand

Peter Robertson
School of Physics
University of Melbourne
Melbourne, VIC, Australia

Woodruff T. Sullivan III
Astronomy Department
University of Washington
Seattle, WA, USA

ISSN 2509-310X   ISSN 2509-3118   (electronic)
Historical & Cultural Astronomy
ISBN 978-3-319-91841-9   ISBN 978-3-319-91843-3   (eBook)
https://doi.org/10.1007/978-3-319-91843-3

Cover image: Chris Christiansen and the solar grating array used to produce radio maps of the Sun. The
array was located at the Potts Hill field station, one of a number of sites in and around Sydney operated
by the Radiophysics radio astronomers. Credit: CSIRO Space and Astronomy

This Springer imprint is published by the registered company Springer Nature Switzerland AG
The registered company address is: Gewerbestrasse 11, 6330 Cham, Switzerland

Australia's history was linked with astronomy from the start – Captain Cook's first voyage was as much to observe the 1769 transit of Venus across the solar disc in Tahiti as it was to explore for *Terra Australis Incognita*, leading to the discovery of Australia's eastern coast the following year. But 175 years would pass before Australians became part of the first rank of world astronomical research. And when this happened, from cliff edges only a few miles removed from Cook's landing site at Botany Bay, it was in a most unlikely manner, for they did their astronomy not with glass lenses, but with rods of metal.

Woody Sullivan, 1988

Foreword

Following the end of the Second World War, scientists and engineers from Allied countries adopted their newly acquired skills and instrumentation to a variety of research programs in basic science. But with growing pressures from the escalating Cold War conflict, military patronage remained an important component of the research enterprise, particularly in the United States. Sparsely populated Australia, with limited resources, could not compete with the big American and European programs in atomic and nuclear physics, and focused its national efforts toward the new science of radio astronomy.

The *Golden Years of Australian Radio Astronomy – An Illustrated History* is an extraordinarily detailed and well documented account of the beginnings of radio astronomy at the Australian CSIRO Radiophysics Laboratory (RP). It describes the semi-independent, and sometimes competitive, pioneering work done at a variety of field stations, mostly located in the Sydney suburbs. Separate research groups were led by Bernie Mills, Chris Christiansen, John Bolton, Frank Kerr, Paul Wild, and others, all under the leadership of Joe Pawsey. None of them had any background in astronomy and, with the exception of Pawsey, most of the Radiophysics scientists and engineers had no postgraduate degrees. However, they not only made major discoveries that changed the course of astronomy, they also developed the new techniques of interferometry and synthesis imaging, the cross-type array later known as the Mills Cross, radio polarimetry, and the solar radiospectrograph, each of which became a basis for radio astronomy research for decades to come. The RP staff all had backgrounds in radio physics, mostly gained from their wartime radar activities. Generally, their lack of familiarity with astronomy was not an impediment to discovery, although there were some stumbling incidents, such as when Bolton, Stanley and Slee argued that M87 and NGC 5128 were Galactic objects, rather than accept the extraordinary radio luminosity implied by their extragalactic origin. Nevertheless, they are credited as reporting the discovery of radio galaxies.

Radio astronomy was a new field of research, and the discoveries came quickly, covering widely divergent topics of the Sun and planets, as well as Galactic and extragalactic astronomy. Among the new research areas were: the demonstration that radio source variability was not intrinsic but due to the ionosphere, the first radio surveys of the southern sky, the separate populations of Galactic and extragalactic sources, the first observations of radio source spectra validating their nonthermal origin, the discovery of the Galactic Centre radio source and the first radio supernova remnants, the discovery of radio galaxies with their unprecedented luminosity, the measurement of the angular size of radio sources, the 21 cm delineation of Galactic spiral arms and galactic kinematics, the relation between solar radio emission and sunspot activity, as well as the classification of solar radio bursts. These all helped to bring Australian science to the global stage.

The astrophysical discoveries at RP were accompanied by major developments in instrumentation and techniques that would provide the seeds for new generations of radio telescopes, in Australia as well as world-wide. During this period, an intense rivalry developed between Radiophysics and the radio astronomy group at Cambridge University in Great Britain. The Cambridge group, which consisted primarily of research students, was led by the brilliant, but perhaps somewhat paranoid and secretive Martin Ryle. There were two components to the Cambridge–Sydney rivalry, one based on the interpretation of radio source counts, and the other, lasting until today, involves credit for the development of what Cambridge called "aperture synthesis" and Sydney called "earth rotation synthesis."

In 1955, Martin Ryle made the remarkable announcement that the Cambridge 2C survey showed a huge excess of weak radio sources when compared to a uniform distribution in space. This meant that at large distances there were either more radio sources or that the sources were more luminous. Either way, this implied dramatic cosmic evolution in stark contrast to the expectation of the popular Steady State theory, promoted by Fred Hoyle and others, which required that the Universe be everywhere the same. But a similar sky survey made by the Sydney group, using the Mills Cross, showed no such excess, and indeed did not even confirm the existence of most of the Cambridge sources. Moreover, as the Australian scientists showed later, the Cambridge analysis was faulty, leading to a bias in estimating the number density of weak radio sources, and the Australians argued that their source counts were consistent with the Steady State cosmology. Modern radio source counts are much closer to the Sydney counts than the Cambridge counts, and it is now recognised that most of the sources in the Cambridge 2C catalogue were spurious. Sydney had much better data but got the wrong answer. Cambridge had bad data and didn't understand the effects of noise and confusion, but apparently got the right answer: We do live in an evolving Universe with a finite age, as dramatically verified by the 1965 discovery of the cosmic microwave background by Arno Penzias and Bob Wilson (a former PhD student of John Bolton).

Martin Ryle later received the 1974 Nobel Prize in Physics for the development of high angular resolution radio imaging using multiple interferometer spacings and the rotation of the Earth, a technique they called "aperture synthesis." Ryle's techniques, which were inspired by Cambridge crystallographers, built on the classic Radiophysics paper by McCready, Pawsey and Payne-Scott, who recognised the Fourier Transform relation between the radio interferometer spatial frequency response and the celestial radio image.

But even before the Cambridge work, Christiansen used his cross array to form two orthogonal fan beams. As the Earth rotated, the two fan beams scanned the Sun at different angles. Fourier transforms of the strip distributions were combined using a two-dimensional transform to obtain an image of the Sun. Was this "earth rotation synthesis" as Christiansen later claimed? Why didn't he use the individual interferometer pairs to obtain his Fourier components as Ryle later did? Perhaps, unlike the Cambridge group, he did not have access to the necessary computing power, although as mentioned in Chapter 1, CSIRAC was then one of the most powerful computers in the world, and was located in the same building as Radiophysics. Or perhaps, as an engineer, Christiansen, like Mills with *his* cross array, was more comfortable with a hardware solution than the software solution so effectively implemented later by Ryle and his colleagues at Cambridge.

There was also a certain level of rivalry within Radiophysics bolstered by the competition for resources and for personal recognition. People did not hesitate to criticise the work of other Radiophysics colleagues – but never in print, as Joe Pawsey's steady hand apparently kept the internal rivalry from getting out of hand.

Golden Years of Australian Radio Astronomy is enhanced with almost 300 historical photographs and line drawings, most of which are taken from the CSIRO Radio Astronomy Image Archive (CRAIA) maintained by the Australia Telescope National Facility. There are also short biographical notes on the many personalities that played a key role in the early Australian radio astronomy programs. Perhaps most remarkable of the many distinguished Radiophysics scientists was Bruce Slee, who started his career in 1945 as a radar operator in Darwin with an independent discovery of solar radio emission, as well as an independent recognition of the sea interferometer effect. Seventy years later, he was still active and had more than 200 publications reporting radio studies of the ionosphere, the Sun, Jupiter, comets, radio stars of all types, radio galaxies and normal galaxies, planetary nebulae, exoplanets, quasars, x-ray sources, pulsars, the Galactic Centre, the interstellar medium, and clusters of galaxies, in addition to many first-hand reports on the history of radio astronomy in Australia. Slee probably had the longest scientific career of any radio astronomer, or perhaps any astronomer, and was equally at ease using filled aperture antennas or interferometry, including VLBI. I was privileged to co-author one paper with Bruce, along with Wayne Orchiston, about the life and career of Gordon Stanley.

The authors of *Golden Years of Australian Radio Astronomy* are well qualified to tell the story. Wayne Orchiston worked at RP from 1961 to 1968, the period described in *Golden Years* as "the changing of the guard", and knew most of the RP scientists and engineers who had remained active during this period. He has published dozens of papers on the history of radio astronomy in Australia, France, India, Japan, New Zealand, and the United States. Peter Robertson was the editor (1980–2001) of the *Australian Journal of Physics*, which published the great majority of RP research papers during the period covered in *Golden Years*. Peter is also the author of *Radio Astronomer – John Bolton and a New Window on the Universe*, about one of the heroes of *Golden Years*. Woody Sullivan, who is the author of *Cosmic Noise: A History of Early Radio Astronomy* and other books that document the early worldwide history of radio astronomy, brings his expertise and extensive historical knowledge to *Golden Years*.

I spent two years as a young Research Officer (the lowest rung on the CSIRO scientific appointment ladder) at RP between 1963 and 1965. I had started my PhD studies under John Bolton, but he left Caltech just as I was starting my thesis research, and I continued with Gordon Stanley as my supervisor. After finishing my PhD at Caltech, I followed Bolton to Australia, where I lived at Parkes and came to know many of the RP scientists and engineers discussed in *Golden Years*. However, by the time I arrived at RP in September 1963, Bernie Mills and Chris Christiansen had already departed for Sydney University, along with some of the key RP technical staff, and Joe Pawsey had died. We were not encouraged to fraternise with the RP deserters, and although their offices were nearby, we had virtually no contact. I met Taffy Bowen only a few times; once, shortly after I arrived in Australia, when John introduced us, and later when he threw me off the Parkes telescope control console in order to have his picture taken at the controls. I did get to know him better when he later served as the Australian Scientific Attaché in Washington.

My time at RP coincided with what the authors of *Golden Years* describe as the second phase of Australian radio astronomy. The Parkes radio telescope, known as the Giant Radio Telescope (GRT), was just coming into full operation. John Bolton had moved to Parkes to take day-to-day control of the operation. The field stations, described in Chapter 2, were mostly gone, and the RP radio astronomy research program had become concentrated around the new Parkes 210 foot (64 m) dish. Everyone had their own research program, or formed small teams. The field station culture of the early years carried over to Parkes, with many of the groups building their own receivers or other instrumentation. Observers, even myself as a junior staff member, did not hesitate to rip apart a piece of suspected malfunctioning equipment in the middle of the night in order to implement a hasty ad-hoc repair.

During those early years, there was no competition for observing time and few external observers. Each group of RP scientists indicated what they wanted to do, and agreed on a division of observing time. Living at Parkes, I could squeeze in extra observing time between other scheduled projects or during holidays when others remained in Sydney to be with their families. However, Bolton was firm in reserving daytimes for maintenance even when no maintenance was planned – until I argued that I wanted to observe Mercury.

As an example of the freewheeling culture at Parkes, I was the assigned observer when news arrived from the US about the detection of the 18 cm spectral line of hydroxyl. Bolton and Gardner worked all day and into the night to convert a 21 cm paramp receiver to 18 cm and then threw me off the telescope so that they could observe OH from the Galactic Centre. After a hasty analysis of their data, they quickly dashed off a letter to *Nature*. I had a perverse satisfaction when it turned out that they had mistakenly interpreted the deep Galactic Centre absorption line as an instrumental baseline error.

The final chapter *Where did it all Lead?* gives a brief summary of the major changes in facilities, sites, and culture that followed the pioneering years after 1960. First came the Parkes 210 foot promoted by Taffy Bowen, then Paul Wild's solar

Radioheliograph at Culgoora, followed by Chris Christiansen's Fleurs Synthesis Telescope, and Bernie Mills' SuperCross at Molonglo. These four projects ushered in the era of "big science" and the demise of the semi-autonomous field station culture. The 1980s saw the Australia Telescope Compact Array become a national facility, the conversion of the SuperCross to the Molonglo Observatory Synthesis Telescope (MOST), and the increasing use of radio astronomy facilities by scientists and students from other than RP or the University of Sydney.

The book concludes with speculations about the future of radio astronomy in Australia and the role of the ambitious Square Kilometre Array project. As the authors remark, although the first germs of the SKA came from Europe, for more than a decade the leadership of this growing international project, the recruitment of international partners, and the guidance of the international governance structure was centred in Australia. With an ideal radio quiet site in a remote part of Western Australia, highly regarded expertise and experience in radio astronomy, particularly in the sophisticated area of synthesis imaging, a well-established national infrastructure, and recognised political, social, and economic stability, Australia seemed the logical choice to build and operate the SKA. South Africa had none of these attributes, except radio quiet sites. But with unreserved enthusiasm and optimism that the international SKA would be built in Australia, CSIRO encouraged the interest and later a proposal from South Africa, thinking this would help promote the visibility and credibility of the increasingly expensive international project.

Two so-called SKA 'pathfinders' (ASKAP – the Australian SKA Pathfinder and MWA – the Murchison Widefield Array) were built in Western Australia, as well as one in South Africa (MeerKat). However, as described by the authors, the final decision to split the SKA between South Africa and Australia was a great disappointment for Australian radio astronomers. Whether or not this surprising and unexpected decision was the result of a concern that, if the SKA were located in Australia, there might be a diminished role for the other countries remains a matter of speculation. With the increasing costs and descoping of the SKA, along with uncertain international funding, ASKAP and the MWA themselves may well be the basis of an exciting future for Australian radio astronomy.

Charlottesville, Virginia   Ken Kellermann

Preface

Thinking back, we should thank Bruce Slee for the appearance of this book. Bruce was one of the pioneers of Australian radio astronomy and was a champion of amateur–professional collaboration in astronomy. When the first author of this book (WO) formed the North Shore Astronomical Society (NSAS) in mid-1959, soon after moving from New Zealand to Sydney, Bruce became a strong supporter of this group of teenage amateur astronomers. Subsequently, one of the projects he devised was for groups of Sydney amateur astronomers to carry out visual monitoring of selected dMe flare stars while he observed them with radio telescopes, searching for radio emission that correlated with optical flares. I led one of these groups, and the Director of Sydney Observatory Dr Harley Wood (who also was a strong supporter of the NSAS) allowed us access to a beautiful old 6-inch refractor that was housed in one of the Observatory's domes.

Then, when I completed secondary school in late November 1961, somehow Bruce Slee had convinced his employers, the CSIRO Division of Radiophysics (henceforth RP) to award me a two-month Vacation Assistant position. It was years later that I discovered that only two of these highly sought after positions were awarded annually, and normally to university students who had already completed MSc degrees and were about to begin their doctoral research. To award one to an 18-year old straight out of secondary school was unheard of.

The two-month position was an amazing experience, and one that was destined to shape the rest of my life and – ultimately – lead to this book. Each day I joined Bruce and two RP technicians and we headed to the Fleurs field station at an old WWII air strip just west of suburban Sydney. My job was to assist Bruce with his work, and at the same time learn about radio astronomy – for as an amateur astronomer committed to optical observing my knowledge of radio telescopes and radio astronomy was negligible (a little like Bruce and his colleagues, who as radio engineers and technicians, had been forced to learn astronomy back in the mid to late 1940s).

I enjoyed that short Fleurs 'apprenticeship' immensely, and came to understand something of the Shain Cross that Bruce, Charlie Higgins and I were using for our observations. But more than this, I learnt about the awe-inspiring discoveries that had been made earlier at Fleurs, and at other RP field stations scattered around suburban Sydney. Dover Heights, Hornsby Valley and Potts Hill all became familiar names, along with Dapto near the industrial city of Wollongong to the south of Sydney. Then all-too-quickly the two-month vacation position came to an end. But I must have made an impression because RP decided to employ me full-time, and they offered me a position as a Technical Assistant. As someone without any academic qualifications I was at the very bottom of the 'employment ladder', but that did not matter – I could continue to work in radio astronomy, and I could learn more about the achievements that by 1962 had made RP a world leader in this newly emerging field of science. The only downside was that I was assigned to the Solar Group, and could only collaborate with Bruce when he needed an assistant during observing sessions on the Parkes Radio Telescope (after the various RP field stations had been closed down). But that did not matter because I also was vitally interested in solar astronomy.

Now let us step forward almost 40 years, to 2000. Armed with a PhD and after an academic career in Melbourne and New Zealand I returned to Sydney where Ron Ekers decided to employ me as Archivist and Historian at the Australia Telescope National Facility (ATNF), the successor to RP. One of my responsibilities was to research the ATNF's remarkable collection of historical images assembled by professional photographers employed by RP from the very early days of Australian radio astronomy. But the photographic collection included more than just radio astronomy, for at the end of WWII RP also experimented with electronic computers, commercial radar systems, vacuum physics, police radars, as well as rain and cloud physics that were to become the Division's other major research forte. My first task was to separate the non-radio astronomy and radio astronomy images, and then catalogue the latter. It was then that the idea emerged to use this incomparable visual record as the basis of a book on the RP field station era. This made good sense as Bruce Slee (by then retired, but continuing his work as an ATNF Research Associate) and I had already started writing papers on the different RP field stations, and some of their unique radio telescopes (such as the Dover Heights 'hole-in-the-ground' antenna). And soon we would start to co-supervise PhD students Harry Wendt and Ron Stewart who respectively would document the main achievements of the Potts Hill and Murraybank, and the Penrith and Dapto, RP field stations.

I then discussed the idea of a copiously-illustrated scholarly book about the RP field stations with Springer, who agreed to be the publisher, but on the understanding that the ATNF would prepare a camera-ready version of the manuscript. I also invited Woody Sullivan to join me as a co-author, so that I could access the files on RP radio astronomy that he had accumulated when researching his classic, *Cosmic Noise – A History of Early Radio Astronomy* (Cambridge University Press, 2009). He agreed, and in 2004 I spent two months in the Astronomy Department at the University of Washington, Seattle, working my way through Woody's incomparable collection (which has since been transferred to the National Radio Astronomy Observatory Archives in Charlottesville, Virginia). Once back in Sydney I commenced writing, and by 2005 I had completed most of the text of the book, selected images for the first four of the five chapters, and prepared captions for them.

Meanwhile, the scheduled fifth and final chapter in the book, about the post-1960 era in Australian radio astronomy, was to be written by the ATNF's Jessica Chapman, as the third co-author of the book, and she also was responsible for preparing the camera-ready manuscript.

Then came a major event that threatened to halt this project altogether or at very least substantially delay publication: Ron Ekers was awarded an inaugural and prestigious Federation Fellowship and stepped down as Director of the ATNF so that he could concentrate on full-time research. Unfortunately, Ron's successor had no sympathy for historical research at the ATNF and suddenly I was out of a job. Meanwhile, Jessica found it increasingly difficult to allocate time for preparation of the camera-ready manuscript of the book, let alone to research and write up Chapter 5. Eventually, Springer's policy also changed, and they were no longer willing to accept camera-ready manuscripts of new books; all production was now to be done in-house. This led to an impasse, and several years passed without any resolution.

Eventually, Jessica left the collaboration, and Woody Sullivan and I signed a new contract with Springer, but there was still a major problem: gaining access to many of the ATNF images required for the book. And Chapter 5 still had to be researched and written. Meanwhile, I was busy writing other books for Springer, and also teaching a History of Astronomy course for the on-line Master of Astronomy degree at James Cook University in Townsville, Queensland (which is where I moved after leaving the ATNF), and supervising further PhD students. One of these students, Peter Robertson, was researching John Bolton's major contribution to international radio astronomy, including his pioneering work with Gordon Stanley and Bruce Slee at the Dover Heights field station in the late 1940s and early 1950s. Peter had also researched the RP field stations when he prepared his book *Beyond Southern Skies – Radio Astronomy and the Parkes Telescope* (Cambridge University Press, 1992), the standard work on the history of this iconic Australian radio telescope. Peter and I agreed that after he submitted his PhD thesis and after his Bolton book was published he would join with Woody and me and complete the long-delayed field stations book. Springer was very relieved to hear this news, even if it meant yet another contract.

Once Peter began updating the original manuscript it soon became obvious that we all had grossly underestimated what this task would entail. A great deal of additional historical research had been published on Australian radio astronomy since 2005 – indeed, it is probably true that this topic has received more attention from historians of science than any other branch of Australian science. All this new material had to be absorbed and assimilated into the text, and the new references added. Peter also agreed to write a detailed version of Chapter 5, an overview of the development of Australian radio astronomy since 1960. The Parkes Telescope, the Molonglo Cross and its evolutionary derivatives, the Fleurs Synthesis Telescope and the Australia Telescope have been discussed in Chapter 5, along with the Square Kilometre Array and the two Australian 'pathfinders'. One outcome of having a Chapter 5 much larger than originally intended was that the book no longer relied almost entirely on the collection of ATNF historical photographs, as we were able to include a much wider range of images. However, a significant setback occurred when the ATNF collection – known as the CSIRO Radio Astronomy Image Archive (CRAIA) – was shut down for almost two years while the database and software underwent a major upgrade, which further delayed completion of the manuscript. Online access to CRAIA was finally restored in September 2019, allowing us at long last to bring this 15-year project to a successful end – much to the great relief of Springer, not to mention all three authors.

This long and difficult project would never have been completed but for the assistance of many people. For their support and encouragement throughout we owe special thanks to Phil Edwards, Ron Ekers, Dick Manchester and Bruce Slee, and to the staff at Springer, especially Maury Solomon and Hannah Kaufman, for making it possible to bring this project to fruition.

For their valued comments and insights on Chapter 5, we thank Bob Frater, Dave Jauncey, Ken Kellermann, Dick Manchester and Richard Schilizzi.

Just over half of the nearly 300 images in this book have been drawn from CRAIA, hosted by the ATNF in the Sydney suburb of Marsfield. For her advice and support we are especially grateful to Jessica Chapman, who has managed the CRAIA project for over two decades and done much of the technical development. We are also grateful to Barnaby Norris who digitised the great majority of CRAIA images and helped set up the original web version of CRAIA; to Lawrence Toomey who sent us high resolution versions of the CRAIA images in this book; and to Vicki Drazenovic who, together with Jessica, did the layout for the first two chapters in the original version of this book (before Springer changed its policy on camera-ready copy).

We would particularly like to thank the following for providing us with one or more photographs reproduced in this book: Letty Bolton, Ellen Bouton, Alex Cherny, Joe Diamond, Martin George, Mary Harris, Lisa Harvey-Smith, Bob Hewitt, Dick Hunstead, Ken Kellermann, Ros Madden, Chris Phillips, Jim Roberts, Bruce Slee, Stephen and Teresa Stanley, Govind Swarup, Harry Wendt and Pete Wheeler. A special thanks to Geoff Kelly who improved the quality of a significant number of images with Photoshop.

We are especially grateful to CSIRO Space and Astronomy for providing a generous publication grant, which has enabled our work to be published in the Springer series of Open Access books.

As this book goes to press, we celebrate the 75th anniversary of the birth of Australian radio astronomy. In October 1945, Joe Pawsey, Lindsay McCready and Ruby Payne-Scott made the first successful radio observations of the Sun from the northern beach suburb of Collaroy in Sydney. We salute them, and all those pioneers who followed and contributed to the remarkable story of Australian radio astronomy. Their fascinating achievements are scattered throughout this book.

Finally, we wish to dedicate our book to Bruce Slee, whose perseverance and achievements have been an inspiration to all of us.

Chiang Mai, Thailand   Wayne Orchiston
Melbourne, Australia   Peter Robertson
Seattle, USA   Woodruff T. Sullivan III

Contents

Chapter 1
From Radar to Radio Astronomy

Today's astronomers study the sky at a wide range of wavelengths, spread across the electromagnetic spectrum, from radio through microwave, infrared, the optical range, the ultraviolet, X-rays and gamma rays (Fig. 1.1). They also use cosmic rays and neutrinos, and the newest field is gravitational wave astronomy. Some of these types of radiation can be observed from the Earth's surface, others rely on space telescopes. Some are comparatively recent innovations, while optical astronomy – in various guises – dates back many millennia.

In radio astronomy we study the radio waves emitted by the Sun, planets and comets of our Solar System, and the stars, dust and gas in our Galaxy and in other galaxies and clusters of galaxies. While the concept of extraterrestrial radio emission dates back to the 1890s, radio astronomy as a new and exciting branch of astronomy only had its origins in the early 1930s when the American physicist Karl Jansky (1905–1950) detected 20 MHz radio emission from the Galaxy (Fig. 1.2; see Sullivan, 2009 for a comprehensive and definitive history of radio astronomy through to 1953). Jansky was a remarkable man (see Biobox 1.1), and it is unfortunate that his employer, the Bell Telephone Laboratories, did not let him pursue these early ground-breaking discoveries further. Instead it was left to Grote Reber (1911–2002; Biobox 1.2) to take radio astronomy to the next level, and to interest optical astronomers in the research potential of this new field. Radio engineering was Reber's profession *and* hobby, and after trying unsuccessfully to obtain a position with Jansky and to interest some leading optical astronomers in 'cosmic static', he realised "Nobody was going to do anything. So … maybe I should do something. So I consulted with myself and decided to build a dish!" (Kellermann, 2005: 49).

© The Author(s) 2021
W. Orchiston et al., *Golden Years of Australian Radio Astronomy*,
Historical & Cultural Astronomy, https://doi.org/10.1007/978-3-319-91843-3_1

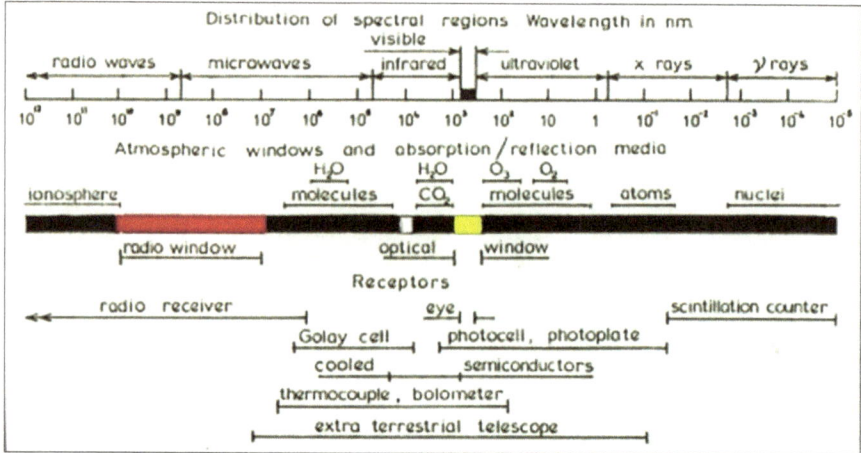

Fig. 1.1 The electromagnetic spectrum, showing the span of the radio window (courtesy: Orchiston collection)

Fig. 1.2 Karl Jansky used the 20.5 MHz 'merry-go-round' antenna at Holmdel, New Jersey, for his studies of extraterrestrial radio emission (courtesy: US National Radio Astronomy Observatory)

In his spare time he built a 9.75 m (32 ft) diameter parabolic dish in the property next door to his mother's home in Wheaton, Illinois (Fig. 1.5). This was in 1937, and although "Neighbours speculated about the purpose of the unfamiliar-looking structure ... Reber's mother found it a convenient place to hang her washing!" (Kellermann, 2005: 49). After experimenting with different receivers, he eventually

Biobox 1.1: Karl Jansky

Karl Guthe Jansky (Fig. 1.3) was born in Oklahoma on 23 October 1905, and graduated with a BSc in physics from the University of Wisconsin in 1927. After a year of post-graduate study, he joined the Bell Telephone Laboratories (BTL), but because he was diagnosed with a kidney disease (Bright's disease) was posted to a rural radio field station at Cliffwood, New Jersey, rather than the main research laboratory in New York. He began researching static associated with shortwave radio transmissions in August 1928, and in 1930 the group transferred to nearby Holmdel. It was there that he built the 'merry-go-round' antenna. In an August 1931 report "… we find what in retrospect appears to be the first recognition of a new, weak component of static." (Sullivan, 1984b: 10). This weak emission, recorded at 20.5 MHz, was researched during 1932 and first mentioned in a research paper by Jansky published that year in the *Proceedings of the Institute of Radio Engineers*.

Jansky continued his investigations on an intermittent basis during 1931 and through 1932 as the Great Depression altered the BTL's research priorities, but by December 1932 he was in a position to conclude that the mysterious 'static' was extraterrestrial in origin. The concept of radio astronomy was born. In 1933 Karl spoke about his research at meetings in Washington and Chicago on 27 April and 27 June respectively. He also published research papers on this work in *Nature* and *Proc. IRE*. Following the Washington meeting, the BTL issued a press release, and on 5 May the front page of the *New York Times* devoted an entire column to the "New radio waves traced to the center of the Milky Way …". Jansky became an instant international celebrity! By August 1933 Jansky was convinced that the static originated from the entire Milky Way, and not just from the central region. The following month Jansky targeted his findings at an astronomical audience through an article in the magazine *Popular Astronomy*, but this was also the time that Jansky turned to other static-related research that was more appropriate to BTL's practical needs.

Fig. 1.3 Karl Jansky (1905–1950) was the founding father of radio astronomy (courtesy: US National Radio Astronomy Observatory)

(continued)

Biobox 1.1 (continued)

Only in 1935 did he briefly return to his 'star noise' research, and publish two final papers in *Proc. IRE* in 1935 and 1937. This marked the end of Jansky's exciting foray into the field that would ultimately be known as radio astronomy, but within months of submitting his 1937 paper the cause would be taken up by another pioneer, Grote Reber, who began building a 9.75 m diameter parabolic radio telescope at Wheaton, Illinois. For his part, Jansky "… continued on the same theme of understanding and minimising the sources of noise in radio communications, whether they were internal to the electronics or external to the antenna, manmade or natural." (Sullivan, 1984b: 24), and his merry-go-round was used as a mount for a variety of different antennas (see also Sullivan, 1984c).

With the passage of the years, Jansky's lifelong kidney ailment led to blood pressure problems, and he eventually died of a stroke in February 1950; he was just 44 years of age. While he never received any official honours or rewards during his short lifetime, other than being made a Fellow of the Institute of Radio Engineers, Karl will always be warmly remembered as the founder of radio astronomy:

> There can be no doubt that he is the father of radio astronomy – but (shifting metaphors) only in the sense of finding and sowing the seed, not in raising the crop. Through a combination of circumstances, his discovery fell on stony ground and was not to yield fruit until the technical demands of a war created a new generation of men and equipment. (Sullivan, 1984b: 35–36).

Biobox 1.2: Grote Reber

Grote Reber (Fig. 1.4) was born in Chicago on 22 December 1911, and obtained a degree in Electrical Engineering from the Armour Institute of Technology (now the Illinois Institute of Technology). From his teens, he was an avid radio ham, and after reading about Jansky's work he decided to conduct research in this new field and in 1937 he built a 9.75 m parabolic transit dish. Subsequently, he published a succession of research papers, mostly on 'Cosmic static'. It is fair to say that initially his work was viewed with skepticism by most optical astronomers (see Reber, 1984 for a personal account).

Soon after WWII, Reber accepted a position at the Central Radio Propagation Laboratory of the National Bureau of Standards and sold his dish and associated instrumentation to his employer on the expectation of building a 23–30 m dish for research. This did not happen, and he quickly became disillusioned and decided to move to Hawaii where he could conduct independent research in radio astronomy. There he erected a sea interferometer on the summit of Mt Haleakala but it produced disappointing results. In November 1954 he moved to Tasmania, and over the next half-century built a succession

(continued)

Fig. 1.4 Grote Reber (1911–2002) laid the foundations of radio astronomy during the late 1930s and early 1940s (courtesy: US National Radio Astronomy Observatory)

of very low frequency arrays and published a series of paper on his research (George et al., 2015b, 2015c, 2017). Whilst widely regarded as one of the 'fathers of radio astronomy', Reber unfortunately promoted cosmological views that were at odds with most other astronomers. He also had little respect for the standard refereeing process used by scientific journals, and was openly critical of politicians, the National Academy of Science and the National Science Foundation in the USA and the work done at the US National Radio Astronomy Observatory. He was able to adopt this stance and maintain his independent status because he received strong financial support over several decades from the Research Foundation.

Although Reber operated outside the 'mainstream astronomical community', he was awarded the Bruce Medal by the Astronomical Society of the Pacific, the Elliot Cresson Medal of the Franklin Institute, and the Russell Lectureship of the American Astronomical Society. He received an honorary DSc degree from Ohio State University, and in 1999 was named 'Man of the Millennium' by the Illinois Institute of Technology. A man of diverse interests, he also conducted research and published on radio-circuitry, ionospheric physics, cosmic rays, meteorology, botany and archaeology. Grote was also deeply concerned about environmental, social and political issues. In Bothwell, Tasmania, he built his modest energy-efficient home, and a battery-powered car. After a long and eventful life, he died in Bothwell in December 2002. For further details of his remarkable life, see Robertson (1986) and Kellermann (2005).

detected Jansky's 'cosmic static' in 1939, and went on to observe the sky at 160 and 408 MHz and to produce the first high-resolution contour maps of Galactic radio noise. Reber wrote papers on his 'cosmic static' and they were quickly published in the *Proceedings of the Institute of Radio Engineers*, but it was a different story when

Fig. 1.5 Grote Reber's
home-made 9.75-metre
parabolic antenna
mystified his neighbours at
Wheaton, Illinois
(courtesy: US National
Radio Astronomy
Observatory)

he tried to share his exciting new finds with astronomers. When he submitted a
paper to the prestigious *Astrophysical Journal* it caused a flurry of excitement and
curiosity. Who was this fellow without an astronomy degree or an observatory affili-
ation, what equipment did he use for these strange observations, and what did they
mean? Several delegations of astronomers, mainly from the famous Yerkes
Observatory, went to Wheaton to meet Reber and see his equipment for themselves.
They seemed suitably impressed, and the editor Otto Struve eventually decided to
accept Reber's paper, but only after some editorial censoring. Reber was less than
impressed with these delays and changes and later stated that while Struve may not
have rejected the paper,

> He merely sat on it until it got mouldy. I got tired of waiting, so I sent some other material
> to the Proceedings of the IRE. It was published promptly ... During the early days of radio
> astronomy, the astronomy community had a poor [publication] track record. The engineer-
> ing fraternity did much better. (Kellermann, 2005: 52).

After making pioneering Milky Way observations, in 1943 Reber went on to
detect radio emission from the Sun. Jansky and Reber, joined by essentially no one
else, can rightly be considered the 'founding fathers' of radio astronomy.

But if radio astronomy was born during the 1930s, it only began to grow during
the late 1940s, mainly because of independent discoveries of solar radio emission in

Fig. 1.6 The RNZAF Wainui radar station near the northern-most tip of the North Island of New Zealand, showing the 200 MHz antenna used for solar observations immediately after sunrise (courtesy: Orchiston collection)

England, the USA, Australia and New Zealand during World War II. Most of these involved radar antennas, and since the 'solar noise' was initially thought to be some form of enemy jamming technique these discoveries were treated as 'secret' or 'top secret'. Strangely, it was the New Zealand discoveries that had the greatest impact in Australia, and led directly to the launch of an ambitious radio astronomy research program by the Radiophysics Laboratory in Sydney (which we shall often refer to simply as RP). The New Zealand solar research was the responsibility of Dr Elizabeth Alexander (1908–1958), Head of the Operational Research Section of the Radio Development Laboratory. Between March and December of 1945 she arranged for solar monitoring to be carried out at a number of different Air Force radar stations (Fig. 1.6). British-born Alexander (see Biobox 3.1 in Chapter 3)

> … prepared a number of reports on this work, and in early 1946 she published a short paper in the newly-launched journal, *Radio & Electronics*. A geologist by training, Elizabeth Alexander happened to be in the right place at the right time, and unwittingly became the first woman in the world to work in the field that would later become known as radio astronomy. (Orchiston, 2005b: 71).

In Australia, Radiophysics had been established at the beginning of WWII in order to develop radar technology for Australian sites and the Pacific theatre (Fig. 1.7). Radiophysics was one of a number of divisions that made up the Council for Scientific and Industrial Research, Australia's leading research organisation. In 1949 CSIR was reconstituted and renamed the Commonwealth Scientific and Industrial Research Organisation (CSIRO) [see Schedvin (1987) and Collis (2002) for official histories of CSIR and CSIRO respectively].

Fig. 1.7 (left) The experimental radar antenna at the Dover Heights field station in Sydney. The antenna was used for the detection of ships along the Australian coast, but then a new version was hurriedly developed for air warning following the Japanese attack on Pearl Harbor. (right) The light-weight/air-warning (LW/AW) portable radar developed by the Radiophysics Lab especially for tropical conditions in the Pacific. The equipment could be airlifted to forward areas and then rapidly assembled. This unit was installed on Tumleo Island near the north coast of New Guinea [courtesy: CSIRO Radio Astronomy Image Archive (henceforth CRAIA)]

In mid-1945 one of Elizabeth Alexander's confidential reports reached the RP Chief, Dr E.G. ('Taffy') Bowen (1911–1991; Biobox 1.3), and his deputy, Dr Joe Pawsey (1908–1962). The New Zealand work fascinated them, and they decided to try and repeat it using a Royal Australian Air Force radar antenna at Collaroy, in suburban Sydney. Pawsey's initial success profoundly influenced the direction of research to be undertaken by Radiophysics in the immediate post-war years, for the Laboratory was under considerable pressure to reinvent itself and find a range of peace-time research projects – any surviving projects with military applications had to be transferred to the defense services.

Biobox 1.3: Taffy Bowen
Edward George Bowen (Fig. 1.8) was born near Swansea in January 1911 and later in life was proudly known by his Welsh nickname of 'Taffy'. As a boy he took a keen interest in radio technology, which sowed the seeds for his future career. He studied physics at the University College of Swansea and in 1934 he was awarded a PhD at King's College London, under the supervision of Edward Appleton. In 1935 Bowen joined Robert Watson Watt's group at the Bawdsey Manor research station working on the highly secret technique of radio detection finding (later known as radar). Taffy was given the responsibility of building the first airborne radar system, which was successfully tested in September 1937. Bowen's other major contribution to the war effort was as a member of the Tizard mission to the United States in 1940 to inform the Americans about recent British

(continued)

Biobox 1.3 (continued)

Fig. 1.8 Dr Edward ('Taffy') Bowen (1911–1991) was Chief of the Radiophysics Lab from 1946 to 1971 (courtesy: CRAIA)

technical advances, including radar. He spent most of the war years shuttling between England and the Massachusetts Institute of Technology which was the centre of US radar research and development (Bowen, 1987).

In late 1943 Taffy accepted an offer to become Assistant Chief of the Radiophysics Lab in Sydney and, two years later, he was elevated to the position of Chief. At the end of the war, the Lab had a group of highly talented staff looking for new research directions. Two major peacetime programs emerged: cloud and rain physics under Bowen's direction, and radio astronomy led by Joe Pawsey. Taffy was a pioneer of cloud seeding experiments in Australia, although his controversial ideas about the influence of meteoric dust on rainfall were mostly discredited. Bowen believed that future advances in radio astronomy would require large and expensive aerial systems and he became the driving force behind the planning and construction of the Parkes Radio Telescope (see Chapter 5). Opened in 1961, the iconic Parkes dish has been at the forefront of astronomical research for over sixty years. The dish was also involved in the series of Apollo missions to the Moon over the period 1969–72. Bowen also made a significant contribution to the establishment of the Anglo-Australian Telescope at Siding Spring in NSW as the inaugural chair of the AAT Board.

In 1972 he was appointed scientific counsellor at the Australian Embassy in Washington, DC, where he spent the remainder of his career. Taffy received a number of prestigious awards and honours over his distinguished career, including the Medal of Freedom USA (1947) for his contributions to radar research and a Commander of the Order of the British Empire (CBE, 1962) for his services to Australian science. In 1975 he was elected a Fellow of the Royal Society of London. For more on Taffy Bowen see Hanbury Brown et al. (1992) and Bhathal (2014).

Fig. 1.9 Testing a
navigational radar system
housed under the 'dome'
on the nose of the aircraft
(courtesy: CRAIA)

Fig. 1.10 A mobile traffic
radar system was
developed by the
Radiophysics Lab for the
New South Wales Police
Force. RP scientist Harry
Minnett (left) explains the
system to the NSW
Minister for Police
(courtesy: CRAIA)

Various fields were investigated in the ten years following WWII, including cloud
physics and rainmaking, navigational aids for civilian aircraft and ships (Fig. 1.9), a
mobile traffic radar system for the NSW police force (Fig. 1.10), and radio astronomy
(or, more properly, 'solar noise' and 'cosmic noise', to use the terminology of the day)
(see Bowen, 1988; Sullivan, 2005). A vacuum physics laboratory was set up, work was
carried out on transistors, and plans were made for a linear accelerator for nuclear physics. Australia's first computer was constructed at this time, known as CSIRAC – the
CSIRO Automatic Computer (Fig. 1.11). CSIRAC was put to work on a number of RP
research programs, but surprisingly had little impact on the radio astronomy program.
Most of the tedious calculations required to reduce observational data were made by
the radio astronomers themselves or by other RP staff (mostly women). Apart from
radio astronomy, the aim of the RP research work was to come up with technological
developments that would assist the Australian economy, and the Division's exploits in
rainmaking (Fig. 1.12) certainly had the potential to do this if they proved successful.

From the start, the RP radio astronomy program was under the capable leadership of Joe Pawsey (Biobox 1.4). Australian by birth, but Cambridge-trained,

Fig. 1.11 Trevor Pearcey and CSIRAC, the first (and last) computer to be wholly designed and built in Australia. CSIRAC is now on display at the ScienceWorks Museum in Melbourne and is the oldest first-generation computer still in existence (courtesy: CRAIA)

Fig. 1.12 During the 1950s and 1960s, apart from radio astronomy, the other major research area at Radiophysics was rainmaking. Silver iodide burners were attached to aircraft and used to seed clouds in selected areas of eastern Australia. Military aircraft were used for the trials because silver iodide is highly inflammable and too dangerous to use on commercial aircraft (courtesy: CRAIA)

Pawsey was a brilliant antenna expert and a real science fanatic, and he simply "… infected everyone with his enthusiasm … He just went from group to group stimulating them, giving them ideas, criticising, and stuff … He was absolutely first class … [and] extremely modest." (Christiansen, 1976). He was seen by most as a respected father figure – even though, for many, he was only ten years their senior – and he was warmly-revered as the "… father of Radio Astronomy in Australia … [and] one of the greatest men in the history of [world] radio astronomy." (Kerr, 1971). While Taffy Bowen was the tough and successful leader of the Radiophysics Lab, Pawsey was the academic scientist and the ideal radio astronomy team leader.

Biobox 1.4: Joe Pawsey

Joseph Lade Pawsey (Fig. 1.13) was born in Ararat, Victoria, on 14 May 1908 and died in Sydney on 30 November 1962. Despite comparatively little formal schooling he won a scholarship to Queen's College at the University of Melbourne, and graduated with BSc and MSc degrees in Natural Philosophy (both with First Class Honours) in 1929 and 1931 respectively. He was then awarded an 1851 Exhibition scholarship, and completed a PhD at Cambridge under Jack Ratcliffe in 1934. Up until the outbreak of WWII he worked for EMI Electronics Ltd, and then returned to Australia and joined the CSIR's Division of Radiophysics in Sydney where he carried out radar-related research.

After the war, Pawsey led the radio astronomy group, formally becoming Assistant Chief of the Radiophysics Division in 1951. He played a pivotal role in most of the important contributions made by this group right up until the time of his decision at the end of 1961 to accept the Directorship of the US National Radio Astronomy Observatory at Green Bank, West Virginia. Joe received the Lyle Medal of the Australian National Research Council in 1953 and the Hughes Medal of the Royal Society in 1960, having been elected a Fellow of the Society in 1954. He was awarded an honorary DSc by the

Fig. 1.13 Dr Joe Pawsey (1908–1962) was the inspirational leader of the radio astronomy group (courtesy: CRAIA)

(continued)

Biobox 1.4 (continued)

Australian National University in 1961, and from 1952 to 1958 was President of Commission 40 (Radio Astronomy) of the International Astronomical Union. Apart from his various research papers, Pawsey was well known for his book *Radio Astronomy* (1955) which was co-authored by one of his disciples, Ron Bracewell. In a 1962 letter to Joe's widow Lenore, the Director of Mount Stromlo Observatory, the inimitable Bart Bok, wrote:

> There are very few scientists in the world who will be able to look back upon a life in which they have helped produce so many distinguished scientists. The young men of Australia who are now the great names of radio astronomy, and who have helped place Australia at the top of the list in the field on a world-wide basis, all express great personal debts for the way in which Joe helped them get started and how he saw to it that their work came to fruition. (Bok, 1962).

In an obituary two of Pawsey's protégés, Chris Christiansen and Bernie Mills (1964), stated that despite his short life,

> It is difficult indeed to over-estimate the value of his contribution to the recent development of the radio sciences and radio astronomy in Australia. Apart from his direct influence in the Radiophysics Division … his influence was felt in the field of optical astronomy, in ionospheric research and in many applications of Radiophysics techniques in other fields.

In 1967 the Australian Academy of Science established the annual Pawsey Medal to recognise outstanding research in physics by an early- or mid-career researcher. For further details of Pawsey's distinguished career see the full-length biography by Goss et al. (2021). See also the Royal Society memoir by Lovell (1964) and for a brief summary Robertson (2000).

Most of the radio astronomy carried out by RP staff between 1946 and 1961 took place at twenty-one different field stations and remote sites in and near Sydney (Fig. 1.14), at two solar eclipse sites in Victoria, two further eclipse sites in Tasmania, and at two temporary field stations in the North Island of New Zealand. Through these,

> … Australia played a key role in the international development of radio astronomy … In addition to standard Yagis and parabolic dishes, innovative new types of instruments were invented, including solar radiospectrographs, solar grating arrays, cross-type radio telescopes, H-line multi-channel receivers, and an assortment of long-baseline interferometers …
>
> Collectively, the radio telescopes at the field stations and associated remote sites were used to address a wide range of research problems, and important contributions were made to solar, Jovian, Galactic and extra-galactic astronomy … largely through the key roles played by leaders such as John Bolton, Chris Christiansen, Frank Kerr, Bernie Mills, Joe Pawsey, Ruby Payne-Scott, Alex Shain, Bruce Slee, Gordon Stanley and Paul Wild. (Orchiston and Slee, 2017: 567).

Rod Davies (1930–2015), who came to Radiophysics in 1950 as a young BSc Honours graduate from the University of Adelaide, described the field station era:

> People tended to explode out first thing in the morning and you didn't see them until the evening. They tended to be quite independent groups [see e.g. Fig. 1.15] and competitive in some ways … Joe Pawsey knew the story well and he would move from one place to another [field station to field station] and keep a very close contact with the various groups (Davies, 1971).

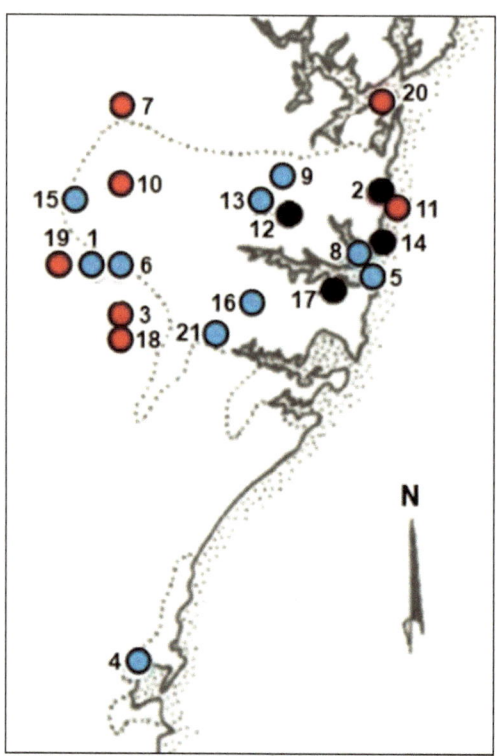

Fig. 1.14 Map showing the location of field stations (in blue), remote sites (red) and other localities (black) in the Sydney and Wollongong regions (indicated by the dotted outlines) that were used for radio astronomy between 1945 and 1961. Key: 1 = Badgery's Creek, 2 = Collaroy, 3 = Cumberland Park, 4 = Dapto, 5 = Dover Heights, 6 = Fleurs, 7 = Freeman's Reach, 8 = Georges Heights, 9 = Hornsby Valley, 10 = Llandilo, 11 = Long Reef, 12 = Marsfield (ATNF Headquarters), 13 = Murraybank, 14 = North Head, 15 = Penrith, 16 = Potts Hill, 17 = Radiophysics Laboratory (grounds of the University of Sydney), 18 = Rossmore, 19 = Wallacia, 20 = West Head, 21 = Bankstown Aerodrome (*map* Wayne Orchiston)

There was a good deal of flexibility about field station life, where staff were free to pursue their own specific research interests (though not all of it was serious science – see Fig. 1.16).

By 1950, the term 'radio astronomy' was meeting gradual acceptance as the generic term for the 'solar noise' and 'cosmic noise' research conducted by the RP staff (indeed it was Pawsey himself who coined the term in a letter to a colleague in January 1948). And yet the name 'astronomy' was in a sense misleading for none of the early pioneers could by any stretch of the imagination initially be classed as 'astronomers'. Indeed, hardly any were former amateur astronomers, and very few had *any* interest whatsoever in astronomy! Rather, most came to 'solar noise' and 'cosmic noise' studies via engineering and physics, although a few had joint engineering/science degrees. Thus the astronomical knowledge of RP staff during the

Fig. 1.15 The Fleurs group in 1956 (from left): Ken Hawkins, Alex Shain, Bruce Slee, Bernie Mills, Kevin Sheridan, Alec Little and Henry Rishbeth (Charlie Higgins was absent at the time). Rishbeth was a visiting PhD student from the Cavendish Laboratory in Cambridge (courtesy: CRAIA)

Fig. 1.16 Bush cricket at the Potts Hill field station with Charlie Fryer (left) and Chris Christiansen (courtesy: CRAIA)

formative years of Australian radio astronomy was almost non-existent, and it is somewhat surprising that Bowen or Pawsey did not devise some formalised means of redressing this shortcoming. Perhaps the fact that many staff were based at the scattered field stations was part of the problem. Instead, it was left to each individual to bone up on astronomy as the need arose. As for basic astronomy references, there were two stalwarts: *Norton's Star Atlas* (which even today remains a favourite with many amateur astronomers), and the two-volume text *Astronomy* by Russell, Dugan and Stewart. But the RP Library did offer other titles, including the Harvard Series in Astronomy, and most of the RP research staff read rapaciously. One of those who sought to bring his overall astronomical knowledge 'up to speed' in a novel way was John Bolton (1922–1993), who methodically read volume after volume of the *Astrophysical Journal* and *Monthly Notices of the Royal Astronomical Society* during lonely nights at the Dover Heights field station when there was nothing significant appearing on the chart records.

Many RP staff also drew widely on the astronomical knowledge of two noted Australian optical astronomers, Claborn ('Cla') Allen (1904–1987) at Mount Stromlo Observatory and the Director of Sydney Observatory, Harley Wood (Fig. 1.17), while distinguished overseas astronomers such as Walter Baade (1893–1960) and Rudolph Minkowski (1895–1976) in California and Jan Oort (1900–1992) in Holland were quick to realise the research potential of radio astronomy and to offer their expertise and their friendship. Following up on these and other contacts, a number of RP staff members attended graduate coursework programs at overseas universities in a bid to rectify their astronomical ignorance. For example, Bernie Mills (1920–2011) went to Caltech in order "… to learn some basic astronomy, which I felt I was sadly in need of in those days. [Also] Caltech was … the obvious place to go to talk to Minkowski and Baade …" (Mills, 1976; see Fig. 1.18). Later Mills (ibid.) was to reminisce:

> From my point of view this was a very fruitful part of my life because I could stop worrying about instrumental developments and sit down and start thinking about astronomy, astrophysics and physics generally … I learned all my astronomy in one year … It put me right in the forefront of thought at that time.

The earliest research at the RP field stations was carried out with simple antennas (e.g. see Fig. 1.19) and with receivers and other equipment that were often left over from WWII, but the international emergence of radio astronomy during the 1950s saw the invention of amazing new types of radio telescopes. These not only opened up exciting new avenues of research, but they also created problems for their

Fig. 1.17 Dr Harley Wood, Director of Sydney Observatory, was a source of astronomical information for the fledgling RP radio astronomers (courtesy: Ros Madden)

Fig. 1.18 Rudolph Minkowski (left) from the Mt Wilson–Palomar Observatories and Bernie Mills. Minkowski visited Radiophysics in March 1956 in an attempt to clear up a controversy between the Radiophysics and Cambridge astronomers (courtesy: *Sydney Morning Herald*)

Fig. 1.19 Simple single, twin and four Yagi antennas, such as this twin Yagi antenna at Dover Heights, were used at the RP field stations in the early days of radio astronomy (courtesy: CRAIA)

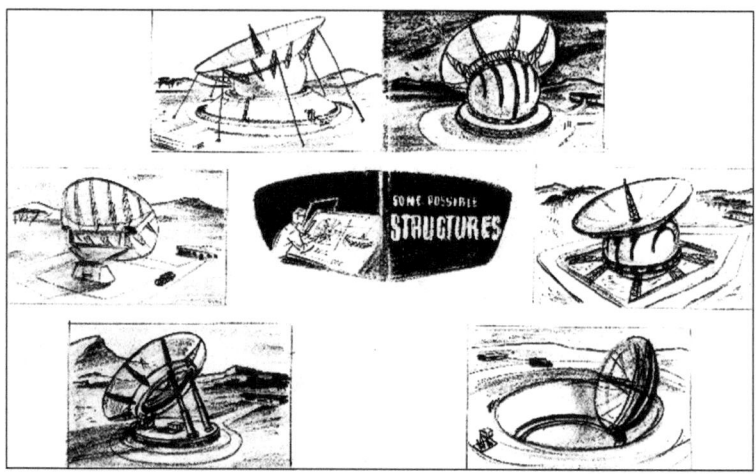

Fig. 1.20 What might have been – some of the designs for a giant radio telescope investigated at Radiophysics in 1954 (courtesy: CRAIA)

inventors as the demand for funds for new instruments began to surpass the funds available in the annual RP budget. In the early 1950s, John Bolton and his Dover Heights team wanted a large new sea interferometer, Jack Piddington (1910–1997) a 6 m diameter dish, Joe Pawsey a large hole-in-the-ground antenna in the Blue Mountains, Bernie Mills a new cross-type radio telescope, and Paul Wild (1923–2008) a suite of solar radiospectrographs. Independent of these proposals that would require internal funding, Taffy Bowen was actively seeking external funds to build a giant radio telescope (Fig. 1.20) that would rival the aperture of the 76 m instrument under construction at Jodrell Bank in England by Bernard Lovell (1913–2012; Fig. 1.21) (see Chapter 5 for details).

After the success of a small pilot model, Mills submitted a proposal for a full-scale cross telescope, which sat on the table alongside the Dover Heights interferometer, Pawsey's hole-in-the-ground antenna, Piddington's dish (which had miraculously grown to about 24 m in diameter), and Wild's new initiative, a solar radioheliograph. Pawsey was placed in the difficult position of having to adjudicate on the competing radio telescope options. If Pawsey had a fault, it was that he hated making 'hard' decisions, and with the passing of the years this was to lead to many problems. In contrast, Bowen was a seasoned campaigner, and "… when Taffy felt strong enough about something, he'd overcome any opposition … Taffy was an enormously energetic person who knew how to get things going on big projects. He knew all the top-level people. He was a *very* good organiser …" (Minnett, 1978).

An immediate loser in this decision-making process was the Dover Heights team, and despite secretly constructing their own large 'hole-in-the-ground' dish-shaped antenna (see Chapter 4), team-leader, John Bolton, soon transferred to the Cloud Physics group within RP (Milne, 1994) and subsequently to the California Institute of Technology where he and Gordon Stanley (1921–2001) established the first major radio astronomy observatory in the United States (Fig. 1.22). By the late 1950s, Radiophysics had "… too many entrepreneurs for the institution to hold

Fig. 1.21 A sketch prepared early in 1952 of the 76 m dish planned by Bernard Lovell at Jodrell Bank in Cheshire (courtesy: Robertson collection)

Fig. 1.22 In 1955 John Bolton and Gordon Stanley from Radiophysics were recruited by Caltech to establish a new radio astronomy group. They founded the Owens Valley Radio Observatory in northern California and designed an interferometer consisting of two 27 m dishes. The dishes were mounted on rail tracks so that the distance between the two could be varied (courtesy: Jim Roberts)

them anymore. It was inevitable that some would move out to other ponds" (Kerr, 1986). As we discuss in Chapter 5, Christiansen (1913–2007) and Mills both lost out when the Culgoora Radioheliograph (Fig. 1.23) was given priority over their own research programs. Consequently, in 1960, both moved to senior posts at the

Fig. 1.23 Sketch of Paul Wild's 96-element solar Radioheliograph (courtesy: CRAIA)

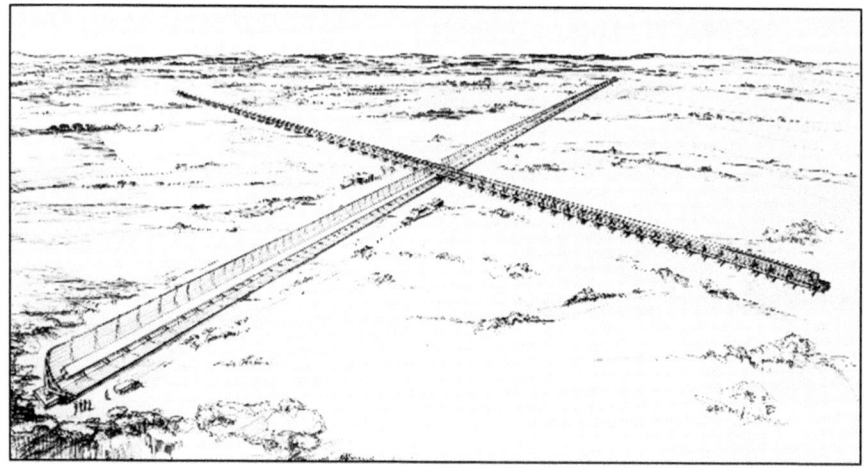

Fig. 1.24 Sketch of the giant Molonglo Cross, planned for Hoskinstown near Canberra (courtesy: CRAIA)

University of Sydney, where Mills went on to build the Molonglo Cross (Fig. 1.24), near Canberra, and Christiansen developed his earlier 'Chris Cross' at Fleurs, in the process converting it into the Fleurs Synthesis Telescope (Fig. 1.25).

Earlier, Ron Bracewell (1921–2007) had transferred to Stanford University in California (Biobox 1.5, Fig. 1.26). Similarly, Rod Davies (1930–2015, Fig. 1.27) joined Bernard Lovell's group at the University of Manchester in 1953 and later became Director of the Jodrell Bank Observatory (1988–1997). Another RP 'graduate' was Kevin Westfold (1921–2001, Fig. 1.27) who moved to the University of Sydney in 1954 and then to Monash University in Melbourne where he eventually became Deputy Vice-Chancellor. We see that as RP's international reputation grew,

Fig. 1.25 View looking west along the E–W arm of the Fleurs Synthesis Telescope. In the foreground is one of the new 13.7 m antennas, and beyond it the smaller dishes belonging to the original Chris Cross (courtesy: CRAIA)

so too did the outflow of major talent, many of them dedicated Pawsey disciples. Even Pawsey eventually joined the exodus, accepting the Directorship of the US National Radio Astronomy Observatory in West Virginia (see Chapter 5).

Biobox 1.5: Ron Bracewell

Ronald Newbold Bracewell (Fig. 1.26) was born in Sydney in July 1921. He was educated at Sydney Boys High and at the University of Sydney where he graduated with the degrees BSc (1941), BEng (1943) and MEng (1948). He then studied for a PhD in ionospheric physics at Cambridge University, under the supervision of Jack Ratcliffe, before joining the Radiophysics Lab in 1949. The physicist Joan Freeman, who was briefly on the RP staff, recalled: "Ron was extremely bright, and had an encyclopaedic range of knowledge: he could discourse eloquently (though sometimes with tongue-in-cheek) on subjects as diverse as metaphysics, ancient history, and cryptography." (Freeman, 1991: 89).

Two examples illustrate Bracewell's various talents. He was appointed secretary of the organising committee for the URSI Congress held in Sydney in August 1952 (see below) and his meticulous planning proved a major factor in the success of the meeting. Although Ron's expertise was in ionospheric physics, Joe Pawsey shrewdly converted him to radio astronomy by inviting him to co-author a book on the subject (Pawsey and Bracewell, 1955).

(continued)

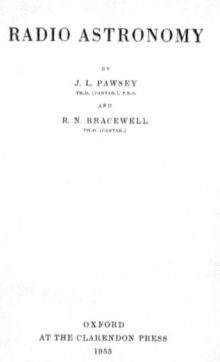

RADIO ASTRONOMY

BY

J. L. PAWSEY
PH.D. (CANTAB.), F.R.S.

AND

R. N. BRACEWELL
PH.D. (CANTAB.)

OXFORD
AT THE CLARENDON PRESS
1955

Fig. 1.26 Ron Bracewell and Joe Pawsey co-authored *Radio Astronomy*, the first scholarly book on the subject. It was published by The Clarendon Press, the imprint of Oxford University (courtesy: NRAO/Associated Universities, Inc. Archives)

In 1955 Bracewell was recruited to establish radio astronomy in the Department of Electrical Engineering at Stanford University, near San Francisco. There he would spend the rest of his career, working well into his eighties. Ron's first project was to build the Stanford Microwave Spectroheliograph, a crossed-grating array consisting of 32 dishes to produce daily radio maps of the Sun (Bracewell, 2005).

The polymath Bracewell was a prolific author of books and research papers. The most important was his book on Fourier transforms and their applications, published in 1965. It was considered the 'bible' on the subject, not only by radio astronomers but also by medical researchers where Ron's mathematical formalism underpinned the development of computer aided tomography (CAT scans). Ron received numerous honours, both astronomical and medical, and in 1998 was made an Officer of the Order of Australia (AO). He died in August 2007. For more on Bracewell's life see Frater et al. (2017) and Thompson and Frater (2010).

Fig. 1.27 Two of the talented young radio astronomers who pursued careers beyond the doors of Radiophysics: Rod Davies (left, University of Manchester) and Kevin Westfold (Monash University) (courtesy: NRAO/Associated Universities, Inc. Archives and CRAIA)

Fig. 1.28 Sir David Rivett (1885–1961) was the inspirational CEO of CSIR during the early post-war years when RP was building its international reputation. He was a strong supporter of radio astronomy and viewed RP as one of CSIR's 'glamour' Divisions (courtesy: CSIRO Corporate Archives)

By the time the Parkes Radio Telescope opened in late 1961, Radiophysics was widely regarded as one of the world's foremost radio astronomical institutions, and a number of people have opined on the range of factors that combined during the late 1940s and through the 1950s to bring about this situation (e.g., see Stewart et al., 2011; Sullivan, 2009). All would agree that having a sizeable highly-skilled radar-orientated workforce that largely stayed together after WWII was a primary ingredient, only made possible by the enlightened view of Sir David Rivett (1886–1961; Fig. 1.28) and the CSIRO Executive towards post-war research in its various Divisions. Another factor was the appointment of a number of first-rate and enthusiastic senior staff members soon after the war; Harry Minnett (1917–2003; 1978), who was later destined to be Chief of the Division, referred to the RP staff of this period as "… a pretty talented bunch of people by any standards." Add to this the initial availability of suitable WWII surplus equipment, an excellently equipped and staffed RP workshop, and the fact that radio astronomy was a new field of endeavour worldwide. It was almost virgin territory and – with some nostalgia – Gordon Stanley (1974) later recalled: "It's the one thing I can never recapture in this game. I mean, the excitement of the whole darn thing at that time … it was so exciting to be in on it … It was simply great fun." (see Fig. 1.29). In a similar vein, Paul Wild (1978) reminisced: "I think we were very lucky in being involved in a very new field. It's so much easier to contribute … [and it] was a marvelously-exciting thing at the time …". As John Murray (b. ca. 1927; 1978) pointed out, "… it was hard to get a non-result …" in this emerging field of science! The lack of any tradition in astrophysics in Australia may also have helped, as it left the early radio astronomers free to 'follows their noses', without being strait-jacketed by optical research priorities.

One reflection of Australia's international standing in radio science generally and in the emerging new field of radio astronomy was the decision to hold the 1952 General Assembly of the International Union of Radio Sciences (known by the acronym of its French name as URSI) in Sydney. This was the first time a major

Fig. 1.29 Gordon Stanley and his Dover Heights receivers in January 1952. Although he went on to become Director of the Owens Valley Radio Observatory in California, Gordon never forgot the nostalgic late 1940s when he, John Bolton and Bruce Slee were among the pioneers of radio astronomy (courtesy: CRAIA)

Fig. 1.30 Upon arrival in Sydney by ship, President of the 1952 URSI Congress, Sir Edward Appleton (centre) is welcomed by Taffy Bowen and Joe Pawsey. At far left is the former RP Chief, Dr David Martyn (courtesy: CRAIA)

international science meeting of any kind had been assigned to a nation outside of Europe or North America (Robinson, 2002), and the decision consigned the non-Australian participants to a lengthy voyage by ship out to what was the far side of the globe. Upon their arrival the President, Sir Edward Appleton (1892–1965), and other dignitaries were greeted with due pomp and ceremony (Fig. 1.30). The conference brought together for the first time many of the world's leading radio astronomers (Fig. 1.31), and combined formal paper sessions with visits to some of the RP field stations (Figs 1.32 and 1.33). It was a great success (see Bolton, 1953; Kerr, 1953).

Fig. 1.31 (above) The URSI delegates outside the Wallace Theatre at the University of Sydney. (below) Some famous faces from the early years of radio astronomy (from left): Chris Christiansen, Graham Smith (England), Paul Wild, Charlie Higgins, Bernie Mills, Jack Piddington, John Hagen (US), Jean-Louis Steinberg (France), Steve Smerd, Jim Hindman, Eric Hill, Alex Shain, Harold 'Doc' Ewen (US), Frank Kerr, Robert Hanbury Brown (England), Lou Davies, Alexander Muller (Netherlands), Ruby Payne-Scott, Alec Little, Marius Laffineur (France), Bruce Slee and John Bolton (courtesy: CRAIA)

All of his contemporaries agree that Australia would not have achieved such international renown without the brilliance, guidance and dedication of one man, namely Joe Pawsey. Initially, Joe believed that there had to be a reasonable compromise between long-term 'mainstream' projects and short-term speculative projects which, if successful, might reveal unexpected exciting new phenomena. By 1954 priorities had changes somewhat, and he felt that while in the past projects were

Fig. 1.32 The Congress
delegates visited a number
of field stations including
Potts Hill where they
examined the grating array
(from left): Chris
Christiansen, Sir Edward
Appleton (UK) and
Balthasar van der Pol
(Netherlands)
(courtesy: CRAIA)

Fig. 1.33 The URSI delegates also visited the Hornsby Valley field station. Robert Hanbury Brown and Graham Smith are fourth and fifth from the left, respectively, while the two RP radio astronomers based at Hornsby Valley, Alex Shain and Charlie Higgins, are third and fifth from the right, respectively (courtesy: CRAIA)

... planned on the basis of a small group building apparatus and using it to get all information possible. We are now moving towards the observatory procedure, where complex
equipment is used by a succession of observers to investigate explicit problems (Pawsey,
1954: 1).

This instrumental constraint was precisely what Christiansen and Mills – amongst others – hoped to avoid. After providing an historical context to his thinking, Christiansen (1986) elaborated:

> The 1950s decade, which I think was by far the best in the history of RP radio astronomy, was not one of transition but one of conflict between two tendencies. One was to continue the invention of new techniques (to beat poverty with brains) and the other was to revert to conventional astronomical observatory practice (of everyone working on one or two large and conventional instruments). Bowen and Bolton represented the latter tendency and Joe [Pawsey] and the rest of us the first. We (Joe etc.) won the first rounds but Bowen won at the end.
>
> Just look at the work of the first half of the decade! There was the radiospectrograph of Wild and McCready, the radio-linked interferometer of Mills and Thomas, my grating interferometers (with resolving power not surpassed in Radiophysics to the present day), and the first maps made by earth-rotation synthesis at Potts Hill, all world firsts, plus the Mills Cross and the interferometry work of Payne-Scott and Little.

Christiansen noted that Pawsey delighted in these developments and felt that it was in the 'true tradition' of the Cavendish Laboratory in Cambridge, where Joe had studied for his PhD. The Cavendish tradition was known as the 'string and sealing wax' approach where major breakthroughs could be made with simple improvised equipment. In other words, for progress in science, brain power was more important than expensive equipment. Christiansen continued:

> Joe was becoming famous while Bowen's own projects were either stalled, or as in the case of his rainfall/meteor shower statistics, treated with derision by statistical experts. Bowen was most impressed by the publicity that Lovell had obtained with the Jodrell Bank telescope and probably felt that if he followed that pattern he could make a real input to radio astronomy which was the only success story in his division of CSIRO. He found an ally in John Bolton who was not amongst the "inventors" and felt that it was time that the group should drop gadgetry and become astronomers. Hence Bowen and Bolton moved for a large simple paraboloid. At first the rest of us looked with favour at this but when it was realised that *this* was to be all that we would have in the future, revolt set in. Most of us had no wish to become conventional astronomers – observers not pioneers.

Pawsey was also the one who facilitated communication between the field stations. Apart from his regular visits, he had staff attend seminars back at the RP Lab, which was situated in the grounds of the University of Sydney in the inner Sydney suburb of Chippendale (Fig. 1.34). These ranged in frequency over the years, from weekly in the early 1950s to perhaps every couple of months in the late 1950s, towards the close of the field station era. Christiansen (1976) fondly remembers them as

> … sort of what the Americans call "bull sessions", thinking of every conceivable sort of aerial and we just about invented every sort possible to invent. It was really good stuff … a really good one [meeting] would last all day. Joe Pawsey was one to stimulate that. He was a marvelous chap.

Staff also received non-intellectual stimulation through social activities, such as parties at the Dapto field station, and there were various avocational outings and specialist-interest group activities (for example, one of the authors (WO) was involved in regular rock-collecting fieldtrips). There was also the Radiophysics cricket team (Fig. 1.35), which started in 1948 and for a number of years participated – with varying degrees of success – in the Sydney inter-club competition.

Fig. 1.34 At the main entrance to the Radiophysics Lab. The administrative, library and workshop staff provided essential support to the RP radio astronomers (courtesy: CRAIA)

Fig. 1.35 The Radiophysics Cricket Team at Chatswood Oval in 1954 (standing from left): John Bolton, Dick McGee, Joe Warburton, Noel Seddon, Alec Weir, Charlie Attwood. (sitting) Charlie Fryer, Paul Wild, Stuart Dryden, Kevin Sheridan, Alan Wightley. Other members of the team, at various times, were Taffy Bowen, Alec Little, Gordon Stanley and Gil Trent (courtesy: CRAIA)

Members of the RP eleven and their supporters practiced religiously at lunchtimes on nearby St Paul's Oval.

Within Australia, during the late 1940s and early 1950s, RP did not quite have a monopoly on radio astronomy. Two small teams of researchers were active in the

Fig. 1.36 (left) Sir Mark Oliphant (Australian National University) was the driving force behind Mount Stromlo's attempts to develop a Department of Radio Astronomy during the early 1950s. (right) Richard Woolley was Director of the Commonwealth Solar Observatory at Mt Stromlo until his appointment in 1955 as Astronomer Royal in England (courtesy: *Life* magazine and Mount Stromlo and Siding Spring Observatories)

immediate post-war years, one at Mount Stromlo Observatory near Canberra and the other at the University of Western Australia in Perth (Orchiston et al., 2006). Their work is discussed in Chapter 3. The Stromlo excursion into radio astronomy was an interesting one and – for a time – promised to threaten RP's supremacy. This only became an issue in 1951 when a meeting of the Observatory's governing board "... reaffirmed the decision that it took a year previously that the radio-astronomy activity of the Observatory ought to be greatly enlarged ..." (Woolley, 1951). This was partly inspired by the fact that the Australian National University (with which Stromlo was newly-affiliated) wanted to offer training in astronomy. Mark Oliphant (1901–2000; Fig. 1.36) was the prime mover at the ANU, and he offered to assist the Observatory in setting up a radio telescope. Observatory Director, Richard Woolley (1906–1986; Fig. 1.36), was expected to draw up a detailed proposal for a Department of Radio Astronomy which would conduct research on Galactic structure, and he looked forward to closely co-operating with Radiophysics.

In February 1951 Woolley announced these plans to Dr Fred White (1905–1994) at CSIRO headquarters (Fig. 1.37). White was Chief Executive Officer of CSIRO and, as a former radio scientist and Chief of RP, he had warm – almost paternal – feelings for the Division, and he immediately discussed the matter with Bowen and Pawsey. The following month, Pawsey and White went to Stromlo where they met with Woolley. In exchange for RP's co-operation Pawsey asked Stromlo to provide optical support for RP's solar radio astronomy program, but Woolley would not agree to this because he intended closing down Stromlo's solar work and redirecting the Observatory's research efforts into Galactic studies. Pawsey and White then suggested that some RP Galactic research could be conducted from Stromlo, to which Woolley agreed.

Fig. 1.37 Sir Fred White
(1905–1994), the charismatic
New Zealand physicist who
came to Australia during WWII
to become Chief of RP. He
subsequently became CSIRO
Chief Executive Officer in 1949
and then Chairman of CSIRO
(1959–70) (courtesy: CSIRO
Corporate Archives)

A major concern for Pawsey was that, even though RP was a world leader in radio astronomy, there had been no involvement of post-graduate students in this research. In the long term, a lack of new talent coming into the field could threaten RP's leadership. He saw Stromlo (through the ANU) as a potential source of students if radio astronomy were developed there. Consequently, as he wrote to Oliphant there "… are obviously good arguments for this and I incline towards it on a long-term view." (Pawsey, 1951). Given Woolley's intention to focus on optical research, Pawsey suggested that RP could establish a radio astronomy branch in Canberra and that at some appropriate time in the future it could break away and join Stromlo.

But the winds of astronomical change were blowing at Mount Stromlo. By the end of 1951 Cla Allen had moved to London and solar research on the mountain was winding down. The Stromlo staff were increasingly preoccupied with the acquisition of a new 74-in telescope (Fig. 1.38), and "Government policy was against any increase in staff … [so] It was impracticable to undertake new work in radio astronomy …" (Notes of a meeting, 1952). Any potential collaboration with Stromlo faded away and RP was left to dominate Australian radio astronomy unchallenged.

Bowen and Pawsey felt this threat from Mount Stromlo keenly, but they were more concerned about RP's standing as a forefront radio astronomy institution, and the threat of international competition. During the decade following WWII, scientists in Australia, Britain, Canada, France, Japan, the Netherlands, the USA and the USSR all made useful studies of 'solar noise' and/or 'cosmic noise'. Australia and Britain quickly became the two leading nations in this new field, and initially relations between all of the radio astronomers were cordial. For instance, in 1948 Pawsey visited Cambridge's Martin Ryle (1918–1984; Fig. 1.39) and discussed the publication of their respective Cygnus A research, along with the results obtained by Bernard Lovell (Fig. 1.40) at Jodrell Bank, as three consecutive letters in *Nature*. Before he returned to Australia, Pawsey also discussed the Cambridge and Sydney solar programs with Graham Smith (b. 1923). Pawsey asked Bowen, to "… arrange [for] a resume of this [RP] work to be sent to him *in an attempt to prevent undesirable duplication.*" (Pawsey, 1948; these are our italics). In contrast, a distinct chilling in relations between the Cambridge and RP radio astronomers – or more precisely between Martin Ryle and Bernie Mills – took place in the mid-1950s

Fig. 1.38 The 74-in Grubb Parsons reflector at Mount Stromlo Observatory near Canberra. Although the telescope was commissioned in 1955, there were 'teething' problems, and it took until 1961 before the telescope was functioning satisfactorily (courtesy: Orchiston collection)

Fig. 1.39 Sir Martin Ryle was head of the Cambridge radio astronomy group and Director of the Mullard Radio Astronomy Observatory (after Hey, 1973: 64). The principal radio telescope by 1958 was an interferometer consisting of a parabolic cylinder reflector 475 m long and 20 m wide, and a smaller mobile reflector which could move along railway tracks (courtesy: CRAIA)

when sources in the Cambridge 2C catalogue were found to be at odds with those discovered by Mills' group at the Fleurs field station (later the 2C catalogue was completely discredited – see Chapter 4 for details). Personalities, credibility and prestige are key ingredients in science, and it took many years before amicable relations were re-established between these two leading research groups.

One of the problems associated with leading a newly-emerging research field from the far side of the globe was the difficulty sometimes encountered in getting credit for discoveries made or new equipment invented. The conventional way to

Fig. 1.40 Sir Bernard
Lovell, head of the radio
astronomy group at the
University of Manchester,
with the 76 m Jodrell Bank
dish in the background
(courtesy: Jodrell Bank
Observatory; all rights
reserved)

bring such developments forward internationally was through publication, but when it came to supposedly reputable journals like *Nature* and even the *Proceedings of the Royal Society*, time and again papers reporting Australian 'firsts' would be held up – sometimes for many months – while English colleagues suddenly came up with the very same findings and were first to rush into print. Sir Edward Appleton, Professor Jack Ratcliffe (1902–1987) and others at Cambridge University were deemed to be the principal culprits (Sullivan, 2009: 90–91 and 110–112). In 1946 Taffy Bowen wrote to Fred White at CSIRO Headquarters in Melbourne following an inexplicable delay in the publication of one of their key papers in *Nature*: "… Appleton is making a song and dance about our letter to *Nature*, but I suppose he is just expressing his well-known "ownership" of all radio and ionospheric work." (Bowen, 1946). Bowen was by no means alone in suspecting Appleton's ultimate motives.

It was partly as a result of this situation, not just at Radiophysics but across the organisation, that CSIRO in 1948 began publication of its own research journal, the *Australian Journal of Scientific Research*, and over the next decade the great majority of research papers penned by RP staff found their way into this journal and its 1953 successor, the *Australian Journal of Physics*. At that time, Australia had no scientific journal devoted solely to professional astronomy, let alone radio astronomy.

To conclude this chapter, let us take a very brief overview of the future development of Australian radio astronomy. Shortly after the Mt Stromlo threat evaporated in 1952, a radio astronomy group was formed at the University of Tasmania (Fig. 1.41), and in 1954 Grote Reber arrived in Tasmania and largely worked as an independent scholar from his home in Bothwell, north of Hobart (Fig. 1.42). For the most part, the Tasmanian work focused on very low frequencies, and as such it complemented – rather than challenged – the RP endeavours (see George et al., 2015a). RP's dominance was not challenged until the 1960s with the advent of the Molonglo Cross (Fig. 1.24) and the Fleurs Synthesis Telescope (Fig. 1.25), constructed and operated by the University of Sydney. Unlike the numerous field stations of the 1950s (the subject of our next chapter), RP operated only two main field stations. The first was the site at Parkes with its famous 64 m radio telescope

Fig. 1.41 The mechanically phase-switched 100 MHz interferometer built around 1952 at Mount Nelson, Hobart, by Gordon Newstead (shown at right) from the Department of Electrical Engineering at the University of Tasmania (courtesy: Martin George)

Fig. 1.42 Grote Reber and the low-frequency array he built at Bothwell in Tasmania. At the time the array had a significantly larger overall collecting area than any other radio telescope in the world (courtesy: Robertson collection and Ken Kellermann)

(Fig. 1.43) that concentrated on Galactic and extragalactic studies (and still does over sixty years later). The other site at Culgoora in northern NSW operated the Radioheliograph (Fig. 1.23) and the radiospectrographs (Fig. 1.44) and concentrated on solar studies. The two University of Sydney instruments, Parkes and Culgoora were the four work-horses of Australian radio astronomy for the 1970s and most of the 1980s. As we will see in Chapter 5, Australian radio astronomy entered a new era with the inauguration of the Australia Telescope in 1988, the year marking the bicentennial of European settlement.

Fig. 1.43 The famous 64 m Parkes Radio Telescope and the 18 m Kennedy dish that originally was sited at the Fleurs field station (courtesy: CRAIA)

Fig. 1.44 The three radiospectrographs at Culgoora eventually spanned the impressive frequency range of 8 to 8000 MHz (courtesy: CRAIA)

References

Bhathal, R., 2014. Bowen, Edward George (1911–1991), *Australian Dictionary of Biography*. Canberra, National Centre of Biography, http://adb.anu.edu.au/biography/bowen-edward-george-17857/text29444.

Bok, B., 1962. Letter to Lenore Pawsey, dated 4 December. Sullivan collection.

Bolton, J.G., 1953. Radio Astronomy at U.R.S.I. *The Observatory*, 73, 23–26.

Bowen, E.G., 1946. Letter to Dr F.W.G. White, dated 26 April. Sullivan collection.

Bowen, E.G., 1987. *Radar Days*. Bristol, Adam Hilger.

Bowen, E.G., 1988. From wartime radar to postwar radio astronomy in Australia. *Journal of Electrical and Electronic Engineering (Australia)*, 8, 1–11.

Bracewell, R.N., 1984. Early work on imaging theory in radio astronomy. In Sullivan, W.T., 1984a. Pp. 167–190.

Bracewell, R.N., 2005. Radio astronomy at Stanford. *Journal of Astronomical History and Heritage*, 8, 75–86.

Christiansen, W.N., 1976. Interview with Woody Sullivan, 27 August.

Christiansen, W.N., 1986. Undated letter to Woody Sullivan, dated about June. Sullivan collection.

Christiansen, W.N., and Mills, B.Y., 1964. Biographical Memoirs [Joseph Lade Pawsey]. *Australian Physicist*, 1, 137–141.

Collis, B., 2002. *Fields of Discovery: Australia's CSIRO*. Sydney, Allen & Unwin.

Davies, R.D., 1971. Interview with David Edge and Michael Mulkay, 5 July. Transcript in Sullivan collection.

Frater, R.H., Goss, W.M., and Wendt, H.W., 2017. *Four Pillars of Radio Astronomy: Mills, Christiansen, Wild, Bracewell*. Cham (Switzerland), Springer International Publishing.

Freeman, J., 1991. *A Passion for Physics: The Story of a Woman Physicist*. Bristol, IOP Publishing.

George, M., Orchiston, W., Slee, B., and Wielebinski, R., 2015a. The history of early low frequency radio astronomy in Australia. 2: Tasmania. *Journal of Astronomical History and Heritage*, 18, 14–22.

George, M., Orchiston, W., Slee, B., and Wielebinski, R., 2015b. The history of early low frequency radio astronomy in Australia. 3: Ellis, Reber and the Cambridge field station near Hobart. *Journal of Astronomical History and Heritage*, 18, 177–189.

George, M., Orchiston, W., Wielebinski, R., and Slee, B., 2015c. The history of early low frequency radio astronomy in Australia. 5: Reber and the Kempton field station near Hobart. *Journal of Astronomical History and Heritage*, 18, 312–324.

George, M., Orchiston, W., and Wielebinski, R., 2017. The history of early low frequency radio astronomy in Australia. 8: Grote Reber and the 'Square Kilometre Array' near Bothwell, Tasmania, in the 1960s and 1970s. *Journal of Astronomical History and Heritage*, 20, 195–210.

Goss, W.M., Ekers, R.D., and Hooker, C., 2021. *From the Sun to the Cosmos, Joseph Lade Pawsey, Founder of Australian Radio Astronomy*. Cham (Switzerland), Springer.

Hanbury Brown, R., Minnett, H.C., and White, F.W.G., 1992. Edward George Bowen 1911–1991. *Historical Records of Australian Science*, 9, 151–66.

Hey, J.S., 1973. *The Evolution of Radio Astronomy*. London, Elek Science.

Kellermann, K.I., 2005. Grote Reber: radio astronomy pioneer. In Orchiston, W., 2005. Pp. 43–70.

Kellermann, K.I., and Sheets, B. (eds), 1984. *Serendipitous Discoveries in Radio Astronomy*. Green Bank (WV), National Radio Astronomy Observatory.

Kerr, F.J., 1953. Radio astronomy at the URSI Assembly. *Sky & Telescope*, 12(3), 59–62, 70.

Kerr, F.J., 1971. Interview with Woody Sullivan, 3 October.

Kerr, F.J., 1986. Letter to Woody Sullivan, dated 12 August. Sullivan collection.

Lovell, A.C.B., 1964. Joseph Lade Pawsey, 1908–1962. *Biographical Memoirs of Fellows of the Royal Society*, 10, 228–243.

Mills, B.Y., 1976. Interview with Woody Sullivan, 25–26 August.

Milne, D.K., 1994. John Bolton and the rainmakers. *Australian Journal of Physics*, 47, 549–553.

Minnett, H.C., 1978. Interview with Woody Sullivan, 2 March.

Murray, J., 1978. Interview with Woody Sullivan, 2 March.

Notes of a meeting held on 7 May 1952 at the Division of Radiophysics, Sydney. Sullivan collection.

Orchiston, W. (ed.), 2005. *The New Astronomy: Opening the Electromagnetic Window and Expanding our View of Planet Earth.* Dordrecht, Springer.

Orchiston, W., and Slee, B., 2017. The early development of Australian radio astronomy: the role of the CSIRO Division of Radiophysics field stations. In Nakamura, T., and Orchiston, W. (eds). *The Emergence of Astrophysics in Asia: Opening a New Window on the Universe.* Springer International Publishing. Pp. 497–578.

Orchiston, W., Slee, B., and Burman, R., 2006. The genesis of solar radio astronomy in Australia. *Journal of Astronomical History and Heritage*, 9, 35–56.

Pawsey, J.L., 1948. Letter to J.A. Ratcliffe, dated 12 November. Sullivan collection.

Pawsey, J.L., 1951. Letter to M.L. Oliphant, dated 10 September. Sullivan collection.

Pawsey, J.L., 1954. Radio Astronomy Group – Current Program. Radiophysics, Sydney. Sullivan collection.

Pawsey, J. L., and Bracewell, R.N., 1955. *Radio Astronomy.* Oxford, The Clarendon Press.

Reber, G., 1984. Radio astronomy between Jansky and Reber. In Kellermann and Sheets (1984). Pp. 71–78.

Robertson, P., 1986. Grote Reber: Last of the lone mavericks. *Search*, 17, 118–121.

Robertson, P., 2000. Joseph Lade Pawsey (1908–1962): founder of Australian radio astronomy. In Ritchie, J. (ed). *Australian Dictionary of Biography*, vol. 15. Melbourne, Melbourne University Press. Pp. 578–579.

Robinson, B., 2002. Recollections of the URSI Tenth General Assembly, Sydney, Australia, 1952. *Radio Science Bulletin*, 300, 22–30.

Schedvin, C.B., 1987. *Shaping Science and Industry: A History of Australia's Council for Scientific and Industrial Research, 1926–49.* Sydney, Allen & Unwin.

Stanley, G., 1974. Interview with Woody Sullivan, 13 June.

Stewart, R., Wendt, H., Orchiston, W., and Slee, B., 2011. A retrospective view of Australian solar radio astronomy 1945 to 1960. In Orchiston, W., Nakamura, T., and Strom, R. (eds). *Highlighting the History of Astronomy in the Asia–Pacific Region: Proceedings of the ICOA-6 Conference.* New York, Springer. Pp. 589–629.

Sullivan, W.T. (ed.), 1984a. *The Early Years of Radio Astronomy. Reflections Fifty Years after Jansky's Discovery.* Cambridge, Cambridge University Press.

Sullivan, W.T., 1984b. Karl Jansky and the discovery of extraterrestrial radio waves. In Sullivan, W.T., 1984a. Pp. 3–42.

Sullivan, W.T., 1984c. Karl Jansky and the beginnings of radio astronomy. In Kellermann, K.I., and Sheets, B., 1984. Pp. 39–56.

Sullivan, W.T., 2005. The beginnings of Australian radio astronomy. *Journal of Astronomical History and Heritage*, 8, 11–32.

Sullivan, W.T., 2009. *Cosmic Noise: A History of Early Radio Astronomy*, Cambridge, Cambridge University Press.

Thompson, A.R., and Frater, R., 2010. Ronald N. Bracewell: an appreciation. *Journal of Astronomical History and Heritage*, 13, 172–178.

Wild, P., 1978. Interview with Woody Sullivan, 3 March.

Woolley, R.v.d. R., 1951. Letter to Dr F. White, dated 12 February. Sullivan collection.

Chapter 2
Frontier Life at the Field Stations

2.1 Introduction

Between October 1945 and March 1946 Radiophysics staff took advantage of the facilities at radar stations located at Collaroy Plateau and Dover Heights in suburban Sydney to follow up the British and New Zealand solar observations mentioned earlier, and this effectively marked the launch of radio astronomy in Australia. The equipment used at these field stations included radar antennas connected to 200 MHz receivers. In 1947 Ruby Payne-Scott (1912–1981) described the Royal Australian Air Force Radar 54 antenna at Collaroy as a

> … broadside array of four horizontal rows each of ten half-wave dipoles with a reflector, having a gain of 80 relative to a half-wave dipole (i.e. g = 130) and a horizontal beam (to the half-power points) of 10°. (Payne-Scott, 1947b)

This COL (chain overseas low-flying) radar at Collaroy and a similar antenna at Dover Heights (see Fig. 2.1) were mounted vertically, and could only rotate in azimuth. In this configuration, they functioned as sea interferometers (Fig. 2.2), and because of the locations of the sites could only observe the Sun as it rose above the eastern horizon.

From early 1946, Radiophysics began setting up radio astronomy field stations in the general Sydney area (see Fig. 1.14), where small groups of scientists and engineers developed new instrumentation and used this in a multi-facetted assault on radio astronomy (see Orchiston and Slee, 2017). One of these pioneers was W.N. ('Chris') Christiansen (1913–2007), who wrote about the field stations:

> The field work had a pioneering appearance. Each morning people set off in open trucks to the field stations where their equipment, mainly salvaged and modified from radar installations, had been installed in ex-army and navy huts. At the field stations the atmosphere was completely informal and egalitarian, with dirty jobs shared by all. Thermionic valves were in frequent need of replacement and old and well-used co-axial connectors were a constant source of trouble. All receivers suffered from drifts in gain and "system-noise" of hundreds or thousands of degrees represented the state of the art. During this period there was no place for observers who were incapable of repairing and maintaining the equipment. (Christiansen, 1984)

© The Author(s) 2021

W. Orchiston et al., *Golden Years of Australian Radio Astronomy*,
Historical & Cultural Astronomy, https://doi.org/10.1007/978-3-319-91843-3_2

Fig. 2.1 (right) The
200 MHz COL Mk V radar
and control room at Collaroy
Plateau in early 1946. The
COL radar was a British
design with the tower
structure built by the NSW
Government Railways (after
Dillett, 2000: 48). (below)
The Dover Heights radar
station in 1943 looking north
towards the entrance to
Sydney Harbour [courtesy:
CSIRO Radio Astronomy
Image Archive (CRAIA)]

Fig. 2.2 Illustrating the
principle of the sea interferometer. The cliff-top aerial
combines the direct signal
from a radio source with the
signal reflected from the sea
to create an interference
pattern. As the radio source
rises above the eastern
horizon, the interference
pattern goes through a series
of maxima and minima. The
reflected signal simulates an
imaginary aerial, spaced from
the real aerial at a distance
equal to twice the height of
the aerial above sea level
(courtesy: CRAIA)

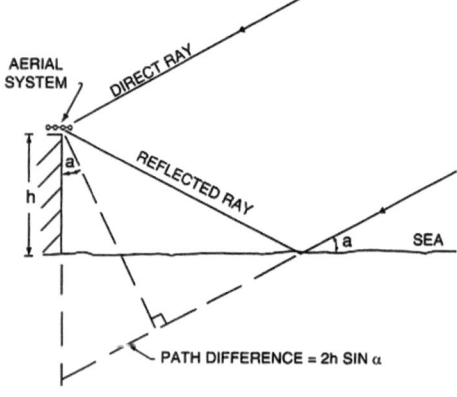

In a document dated 12 October 1956, the head of the Radio Astronomy group at RP, Joe Pawsey (1908–1962), outlined his philosophy regarding field station research:

A major point of policy in directing the Group over the past years has been that the out-standing individual research workers should have their own distinct fields of interest and should have facilities appropriate to the work. In the event of one failing to make good progress ... facilities should not be taken from the others to bolster him up. (Pawsey, 1956)

Meanwhile, a letter written to a US colleague in 1960 reveals more about Pawsey's thoughts on effective research: "So my advice [to you] is to encourage radical ideas and not be afraid of a "damn fool" experiment, provided you can do it quickly and get rid of it quickly if it does not work." (Pawsey, 1960a). But, as Jim Hindman (1919–1999; 1978) has reminded us, the chance of success at the RP field stations was high, for "In those days, you could pick almost any field [of radio astronomy] and make a discovery." Yet Frank Gardner (1924–2002; 1986) was firmly convinced that without Pawsey's influence radio astronomy would have died at Radiophysics: "Personal hostilities and lack of appreciation of fundamentals would have led to chaos, and within a few years the Executive would have encouraged research in radio astronomy, an anomaly within CSIRO, to taper off." Thus, Pawsey played a key role in the success of the research and in the viability of the field stations.

Those of us (like the first author of this book) who were lucky enough to have lived through the 'field station era' remember the field stations with genuine affection. There was a freedom not experienced by those back at the 'Lab' (as the Radiophysics Laboratory was known): the pervading sunshine, the clean fresh air, those incident-packed trips from home to field station (during the 1950s Commonwealth cars replaced the open trucks mentioned above), and the sense that we were somehow making history. There were also snakes to contend with, wet days when antennas still had to be aligned and observations made, floods that had to be negotiated, and those times – fortunately few and far between – when vehicles became bogged and had to be rescued by local farmers (Fig. 2.3). Brunsviga calculators (Fig. 2.4) and slide rules were the norm, and personal computers but a future dream. Signal generators, not sources, provided calibrations, and results were displayed in real time on Esterline Angus and other all-too-familiar chart recorders (Fig. 2.5). These were pioneering days!

At one time or another between 1946 and 1961, ten different field stations existed in the Sydney area and near Wollongong (Figs 2.6 and 2.7), and the development of each of these is discussed in this chapter. In addition, the Collaroy and North Head radar stations were used briefly for solar radio astronomy projects in 1945–46, and during the 1950s and 1960s a number of short-lived 'remote' sites were used in conjunction with the regular field stations (e.g., see Research Activities ..., 1960). In all, research by RP staff was carried out at 21 different sites in the general Newcastle–Sydney–Wollongong area during the early days of Australian radio astronomy, as well as at two solar eclipse sites in Victoria, two solar eclipse sites in Tasmania (see Fig. 2.8) and two temporary field stations on the North Island of New Zealand. In addition, during the 1940s RP staff were closely associated with the

Fig. 2.3 Life on the field stations sometimes had its downsides. A farmer comes to the aid of the mobile lab, affectionately known as 'Flo', and four of the RP staff: Bruce Slee (left), Alec Little, Les Clague and Kevin Sheridan. The poles form part of the Shain Cross, one of the major radio telescopes of the 1950s (see Fig. 2.63) (courtesy: CRAIA)

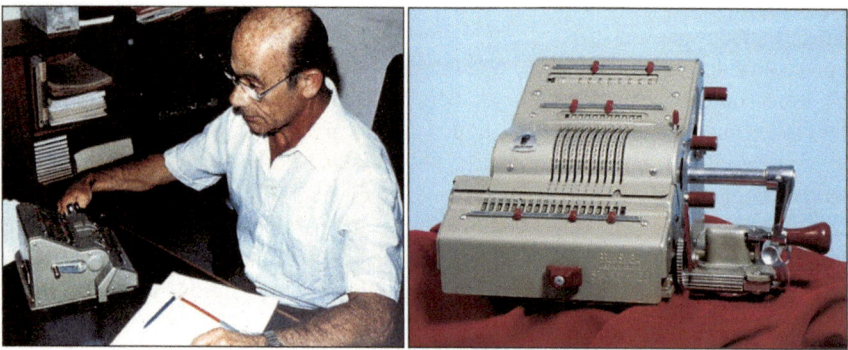

Fig. 2.4 Solar radio astronomer Steve Smerd with one of the Brunsviga calculators (courtesy: Sullivan collection and CRAIA)

solar and Galactic radio astronomy that was conducted at Mt Stromlo Observatory near Canberra.

Staff based at the ten 'regular' RP field stations often enjoyed unheralded visits from Joe Pawsey (Fig. 1.13), head of the Radio Astronomy Group within the Division, the undisputed "… father of radio astronomy in Australia …" (Kerr, 1971). Joe "… had a remarkable physical insight and an almost boyish enthusiasm for just 'suck it and see' sort of attitude, and this really was the key thing in the whole early days of radio astronomy." (Lehany, 1978). He liked to turn up at field

Fig. 2.5 Dick Mullaly at the Fleurs field station examines the chart recorder as the Chris Cross generates a strip scan of the Sun (courtesy: CRAIA)

Fig. 2.6 The locations of ten RP field stations in and near Sydney and Wollongong discussed in this chapter (see also Fig. 1.14): 1 = Badgery's Creek, 2 = Dover Heights, 3 = Fleurs, 4 = Georges Heights, 5 = Hornsby Valley, 6 = Murraybank, 7 = Penrith, 8 = Potts Hill, 9 = Dapto, 10 = Bankstown Aerodrome (*map* Wayne Orchiston)

stations at morning or afternoon tea time, armed with a supply of 'lamingtons', cubes of sponge cake coated in chocolate icing and desiccated coconut, which he found totally irresistible! Frank Gardner (1973) remembers how Joe was "… very good at getting the best out of people. He had this method whereby when you came to talk to him about some problem, he'd often propose some other way of doing it … he took an interest in everything …". The downside of Joe's unbridled enthusiasm was that "… you were lucky to get home for dinner when he showed up at the end of the day." (Payne-Scott, 1978). Yet through these flying visits, Joe was able to keep track of progress at the field stations, and provide feedback to their staff when there were problems to be discussed.

Fig. 2.7 Sketch map of the Sydney area showing the locations of five of the ten RP field stations in Fig. 2.6. The airports at Kingsford Smith and Richmond were used for the rainmaking experiments and for radio applications to air navigation (courtesy: CRAIA)

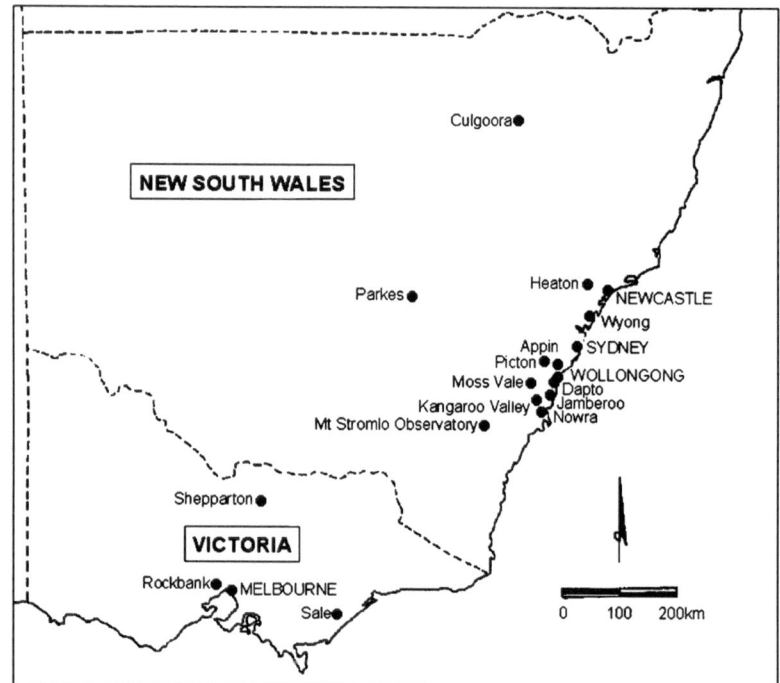

Fig. 2.8 The locations of some other radio astronomy sites mentioned in the text (*map* Wayne Orchiston)

Field station staff mainly heard of developments at other field stations only when they attended occasional seminars and meetings at Radiophysics. Their frequency varied considerably during the early years:

> There were times when these meetings were held very regularly and other times when they seemed to be forgotten altogether … [When they were held] it was a time when one sorted out people treading on each other's toes or pressing each other's borders … (Wild, 1978).

At these times field station staff also could access the Lab's excellent research and reference library, and liaise with the engineers and technicians involved in designing, developing, and overseeing the fabrication of new antennas and receivers. Later, they would make occasional use of the computing facilities offered by the Lab, which built Australia's first home-grown computer and which was appropriately christened CSIRAC (Fig. 1.11).

While most of the early observational work was carried out at the field stations, in 1948 and 1949 some important research was carried out using a small antenna mounted on the 'Eagle's Nest', a small room and associated flat-roofed area located at the very top of the RP Lab. The 1.1 m (44 inch) equatorially-mounted dish – a recycled WWII searchlight mirror – was used by Jack Piddington (1910–1997) and Harry Minnett (1917–2003) in mid 1948 for observations of the Sun and the Moon at 24 GHz (Fig. 2.9). Then between October 1948 and March 1949, Harry Minnett and Norman Labrum (1921–2011) used this same antenna for solar monitoring at 9.4 GHz. A slight interruption to this program occurred on 1 November 1948 when they used the small dish to observe a partial solar eclipse, in collaboration with RP colleagues at the Potts Hill field station and at remote sites in Victoria and Tasmania. This dish was then transferred to Potts Hill.

Fig. 2.9 The 'Eagle's Nest' was a small flat-roofed room at the very top of the Radiophysics Lab that was used for research and to test new equipment. Piddington and Minnett used the 1.1 m antenna to observe the Sun and Moon in 1948 (courtesy: CRAIA)

Fig. 2.10 Graham Smith
with one of the 7.5 m
diameter Würzburg dishes
used in the interferometer
at Cambridge for accurate
measurements of radio
source positions (courtesy:
Robertson collection)

It is interesting to note that Radiophysics made an attempt to obtain one of the
large Würzburg radar dishes, abandoned by the retreating German forces after the
D-Day invasion in June 1944 (see Fig. 2.10). These dishes were fundamental to the
early post-war development of radio astronomy in a number of countries, particularly in Europe (e.g. see Orchiston et al., 2007; Radhakrishnan, 2006; Sullivan,
2009: 78; Van Woerden and Strom, 2006). In November 1945 John N. Briton (ca.
1908–1965), then the acting Chief of the RP Lab, wrote to a colleague in London:

> "We understand that there is a good possibility of sidetracking one of the German Würzburg
> equipments from the Royal Air Force. We would be very glad indeed to acquire one of
> these. We would set it up at our new field testing site at Georges Heights, Sydney, where it
> would be very useful for a number of purposes ..." (Briton, 1945).

Unfortunately, nothing came of the idea.

2.2 Dover Heights

The Dover Heights field station was at the site of a WWII Australian Army radar
station, on the cliff edge some 79 m above the Tasman Sea and 5 km south of the
entrance to Sydney Harbour (Figs 2.1 and 2.6). In the immediate post-war years, it
was undoubtedly the most significant of all RP field stations. It was here that Ruby
Payne-Scott (see Biobox 2.1) became known internationally for her solar work and
John Bolton (1922–1993; Biobox 2.2), Gordon Stanley (1921–2001) and Bruce
Slee (1924–2016) for their pioneering studies of the enigmatic 'radio stars'. It was
also here that Dick McGee (1921–2012) would conduct his first research project,
ideal training for his future involvement in hydrogen-line astronomy at Potts Hill,
Murraybank and Parkes (Sim, 2013).

Biobox 2.1: Ruby Payne-Scott

Ruby Violet Payne-Scott (Fig. 2.11) was born in Grafton, New South Wales, on 28 May 1912 and died in Sydney on 25 May 1981. She graduated from the University of Sydney with a BSc (First Class Honours) in Physics and Mathematics and an MSc in 1933 and 1936 respectively, one of very few women to study science at this time, and in 1938 was awarded a Diploma of Education. From 1936 to 1939 she worked as a physicist for the Cancer Research Department at the University of Sydney and for a year as Science Mistress at Woodlands Church of England Girls Grammar School in Glenelg, South Australia. Between 1939 and 1941 she was a radio engineer at Amalgamated Wireless Australasia, before joining the Radiophysics Lab.

Ruby worked on wartime radar technology at Radiophysics and then carried out pioneering solar radio astronomy at Dover Heights, Hornsby Valley and Potts Hill before leaving CSIRO in 1951 to start a family. Because Commonwealth regulations prevented the permanent employment of married women, she had kept secret her marriage in 1944 to Bill Hall from the CSIRO hierarchy during her years at Radiophysics. Ruby enjoyed one brief return to the astronomical fold when she attended the 1952 URSI Congress, held in Sydney, and she is conspicuous in Fig. 1.31 as the sole female radio astronomer present.

After raising two children, Ruby took up teaching again in 1963, and over the next 13 years successive classes of girls at Danebank Anglican School for Girls in Sydney received their initial training in physics from her. Ruby held strong left-wing political views, was a champion of women's rights, and a keen bush-walker. One of her co-workers, Gordon Stanley (1997), later reminisced: "She was part of my early education on women's issues, and despite early insensitivities on my part, I grew to have a great respect and liking for her."

In 2008 CSIRO established the Payne-Scott Award to support researchers who take extended career breaks to be the primary family carer. In 2021 the Australian Academy of Science established the Ruby Payne-Scott Medal, recognising outstanding career achievements by a woman in the physical or biological sciences. See the full-length biography of Ruby by Goss and McGee (2009) and the abridged version by Goss (2013).

Fig. 2.11 Ruby Payne-Scott (1912–1981) was Australia's first female radio astronomer (courtesy: Bill Hall family; all rights reserved)

Biobox 2.2: John Bolton

John Gatenby Bolton (Fig. 2.12) was born in Sheffield, England, on 5 June 1922. After completing a BA degree at Cambridge University he served as a radio officer on the aircraft carrier HMS *Unicorn*, which was stationed in Sydney at the end of WWII. John stayed on and in September 1946 was appointed a research officer in the Radiophysics Lab. Initially he worked closely with Gordon Stanley and Bruce Slee at Dover Heights, where he "… isolated the first known discrete radio sources, measured their positions and radio spectra, placed upper limits on their angular sizes, recognised the non-thermal nature of the radio emission, and made the first identifications … of radio sources with optical counterparts." (Kellermann, 1996: 729). The discovery of these discrete radio sources marked the beginning of a new branch of astronomy – extragalactic radio astronomy.

Bolton then transferred briefly to the Lab's rainmaking group before accepting a fellowship at the California Institute of Technology in 1955. With colleague Gordon Stanley he established the Owens Valley Radio Observatory (see Fig. 1.22) in the northern Californian mountains, the first major observatory for radio astronomy in the US. At the end of 1960 John returned to Australia as Director of the new Parkes Radio Telescope, where he and his many colleagues carried out two extensive surveys of the southern sky, first at 408 MHz (over 2100 radio sources) and then at 2700 MHz (over 8000 sources). Bolton's principal role was to make optical identifications of these sources, most of which were either radio galaxies or quasars, and this involved extended visits to optical observatories in the US.

Bolton was the PhD supervisor and mentor to many of the next generation of radio astronomers. He was a Fellow of the Australian Academy of Science (1969) and of the Royal Society of London (1973). Among the many honours John received were the Inaugural Karl Jansky Lectureship (1967, US National Radio Astronomy Observatory), the Gold Medal of the Royal Astronomical Society (1977) and the Bruce Medal (1988, Astronomical Society of the Pacific). Following a heart attack in 1979, John retired to the coastal town of Buderim in Queensland, where he died on 6 July 1993. See Robertson (2017) for a full-length biography; see also Wild and Radhakrishnan (1995), Kellermann (1996) and Orchiston and Kellermann (2008).

Fig. 2.12 John Bolton (1922–1993) and the group at the Dover Heights field station discovered the first discrete Galactic and extragalactic radio sources (courtesy: Letty Bolton)

Fig. 2.13 The 4-Yagi antenna installed at Dover Heights in 1946 and operated at 200 MHz. Note the simple equatorial mounting that allowed the antenna to track the Sun and other celestial objects (courtesy: CRAIA)

Payne-Scott's initial solar work at Dover Heights had been carried out with the 200 MHz WWII radar antenna (see Fig. 1.19), but this was not designed for radio astronomy. All it could do was record the Sun as it rose above the horizon; it could not track the Sun throughout the day. This rather serious problem was overcome in early 1946 when RP staff installed a 4-element 200 MHz Yagi antenna (Fig. 2.13) on the roof of the second blockhouse at Dover Heights, about 50 m north–west of the cliff-side blockhouse with the radar antenna. At about the same time, similar 200 MHz Yagi antennas were also installed at the North Head radar station (at the northern entrance to Sydney Harbour) and at Mount Stromlo Observatory, near Canberra.

The first Dover Heights Yagi antenna was soon followed by twin-Yagis operating at 60 and 85 MHz, attached to improved receivers built by Stanley, and in November 1946 a 100 MHz twin-Yagi replaced the 85 MHz antenna. By May 1947, the three Yagi antennas had been transferred to the roof of the seaward blockhouse (see Fig. 2.14). Bolton (1986) later recounted in a letter to one of us (WS) that by February 1947 the rusting old WWII radar antenna "… had been almost destroyed by vandals and only the basic framework was left … [so] Stanley and I cut it up with an oxy torch and dropped the bits over the cliff …"!

Although the simple Yagi antennas were designed initially for solar research, they were soon turned to good account in the search for 'radio stars'. As part of this project, between late 1947 and early 1948, Bolton, Stanley and Slee conducted what was probably the world's first spaced-antenna experiment, using the Yagi aerials at Dover Heights and similar antennas set up at Long Reef and briefly at West Head, about

Fig. 2.14 John Bolton and
the cliff-side blockhouse at
Dover Heights in May
1947, with the 60 and
100 MHz twin-Yagis on
the roof. The 200 MHz
antenna was on the far
corner and is not visible.
Note the wartime
camouflage paint
(courtesy: Stanley family)

15 km and 35 km away and to the north of Sydney Harbour, to simultaneously measure
the intensity variations exhibited by the source in the Cygnus constellation. A further
extension of the radio stars project occurred in 1948 when the Dover Heights 100 MHz
antenna was used in conjunction with a 4-Yagi mobile radio telescope (Fig. 2.15) that
was transported to New Zealand and between June and August was used at two different sites near Auckland (Fig. 2.16) (Orchiston, 1994). Pakiri Hill, near Leigh, on the
east coast, allowed observations of sources as they rose above the horizon, while the
cliffs above Piha, a famous surf beach to the west of Auckland, permitted observations
of these same sources as they set below the western horizon. The cliffs at both Pakiri
Hill and Piha were about three times higher than at Dover Heights, which meant that
the resolution of their sea interferometer aerial was three times better. While Bolton
and Stanley took care of these observations at RP's most distant temporary field stations, Slee continued to conduct parallel observations at Dover Heights.

In order to improve the instrumentation at Dover Heights, in early 1949 a 9-Yagi
array was constructed by RP workshop staff and mounted on the blockhouse. This
had a novel mounting which allowed it to operate as an equatorial (permitting easy
tracking of objects as they moved across the sky), or as an alt-azimuth instrument so
that it could function as a traditional sea interferometer and record sources rising
over the eastern horizon (see Fig. 2.17). It was Stanley who suggested introducing a

Fig. 2.15 The ex-Army radar trailer in the grounds of the Radiophysics Lab, shortly before it was shipped to Auckland, New Zealand. This mobile sea interferometer featured four Yagi aerials, a new 100 MHz receiver, recorders, chronometers, weather recording instruments and all the tools and backup equipment needed to operate reliably at a remote location (courtesy: CRAIA)

Fig. 2.16 The locations of Pakiri Hill and Piha in relation to the city of Auckland. The observations carried out at these two temporary field stations in the winter of 1948 marked the start of non-solar radio astronomy in New Zealand (after Orchiston, 1994: 542)

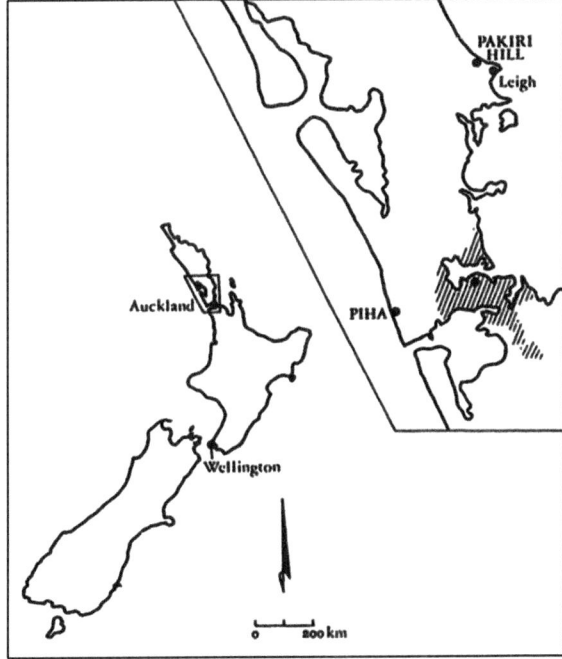

Fig. 2.17 The 9-Yagi array under construction early in 1949, with two of the Yagis leaning against the blockhouse. The array with its equatorial mount was used to carry out the first survey of the southern sky (courtesy: Stanley family)

third axis that in effect converted the declination axis to an azimuth axis for sea interferometry. Bolton and new recruit Kevin Westfold (1921–2001) used this antenna with a 100 MHz receiver to carry out the first systematic survey of the southern sky. At the same time, Stanley and Slee temporarily installed new 2-element Yagis operating at 40 and 160 MHz on the roof of the blockhouse, in order to examine a number of strong discrete sources over a wider range of frequencies.

In 1950 Stanley was largely responsible for constructing a new radio telescope at Dover Heights, a 3.9 m (16 ft) parabolic antenna. Although dishes had been used for some time at the RP Lab and at the Potts Hill field station, this was the first dish to be used at Dover Heights. Initially it was mounted on the back of an old WWII trailer and used as a sea interferometer (Fig. 2.18), but eventually it was attached to the equatorial mounting on the blockhouse roof used formerly by the 9-Yagi antenna (Fig. 2.19). This dish was intended mainly as a test bed for low-noise receivers between 190 and 400 MHz, although it was also used to study some of the stronger discrete sources.

With the dish installed on the roof of the blockhouse, in mid-1951 Bolton, Stanley and Slee began constructing another radio telescope. By using elements of the now-surplus 9-Yagi antenna, they soon assembled near the cliff edge a new sea interferometer that was set up on an azimuthal mounting. This was briefly an 8-Yagi array (Fig. 2.20), but was soon converted to a 12-Yagi array, arranged in two widely-spaced banks of six Yagis (see Fig. 2.21), and was used for yet another all-sky source survey at 100 MHz.

Fig. 2.18   The 3.9 m dish mounted in sea-interferometer mode on the back of an ex-Army truck (courtesy: CRAIA)

Fig. 2.19   Star trails and the 3.9 m parabolic dish on the roof of the Dover Heights blockhouse (courtesy: CRAIA)

The last radio telescope to be built at Dover Heights was an ingenious design, and used typical Australian initiative to overcome funding shortfalls at RP at this time (Orchiston and Slee, 2002a). It represents what Christiansen (1986) referred to as the tendency "… to beat poverty with brains …". By 1951 several different RP staff members were competing for the limited funds available in order to construct innovative new radio telescopes. Bolton, Stanley and Slee wanted to build a very large two-element interferometer at Dover Heights but were unsuccessful in their funding bid, so they decided to take the initiative and build a giant new radio telescope themselves. The idea for the telescope most probably came from Bolton who

Fig. 2.20 The 8-Yagi antenna was built at the end of 1951 and installed on an old wartime mount a little to the south of the blockhouse and close to the cliff edge (courtesy: CRAIA)

Fig. 2.21 The 12-Yagi array, looking south, was used at 100 MHz to carry out a sky survey and detected 104 discrete sources (see Table 4.3) (courtesy: CRAIA)

had spent most of 1950 touring radio astronomy centres in Europe and North America. He visited the Jodrell Bank field station in Cheshire and was impressed by a fixed above-ground parabolic dish, consisting of a spider's web of wires suspended on upright tubular supports. Its huge size (67 m diameter) enabled Robert Hanbury Brown and Cyril Hazard to detect faint radio emission from the giant Andromeda galaxy.

The Dover Heights group realised they could build a similar dish by excavating a dish-shaped depression in the sand and furnishing this with a reflecting surface and a long mast supporting a dipole. Despite the dish being stationary, different sections of the sky could be surveyed by using the Earth's rotation and by altering the angle of tilt of the antenna mast. Sydney's latitude was ideal, for it meant that the

Galactic Centre – a region of great interest to astronomers – would pass nearly over-head. They chose a site about 150 m to the north of the blockhouse, and during a three-month period in 1951 spent their lunch breaks shifting enough sand to create a 22 m paraboloidal shape, dumping the spoil around the rim of the depression. When the hole neared completion, Stanley drove a truck to the Bunnerong Power Station and collected several loads of ash, which were worked into the sandy surface to help stabilise the shape. He then collected many metres of steel strip from pack-ing cases at the shipping terminal in Botany Bay, and these were laid in close paral-lel bands across the depression to form a crude reflecting surface. A guyed aluminium mast with a 160 MHz dipole at its top was erected at the centre of the dish, and a low-noise preamplifier was installed in a waterproof box at the base of the mast. From there, the signal was fed to a nearby mobile laboratory (Fig. 2.22).

Initial observations proved successful and in 1952, with Pawsey's blessing, funds were assigned for an upgrade (Fig. 2.23). Over the next three months the dish diam-eter was extended to 24.4 m, the surface was coated with a layer of concrete in which a small-mesh chicken wire reflecting surface was imbedded (Fig. 2.24). The outer part of the antenna consisted of a 'skirt' formed out of 25 mm aluminium

Fig. 2.22 The original 22 m 'hole-in-ground antenna', one of the largest radio telescopes in the world, was built by the Dover Heights staff as a lunchtime project in 1951. The catwalk provided access to the centre of the dish, the aerial mast and dipole, and the instrument box at the base of the mast that contained part of the 160 MHz receiver (courtesy: CRAIA)

Fig. 2.23 Upgrading the hole-in-the-ground in February 1953 – a concrete surface was added and the diameter extended from 22 to 24.4 m. The rotating parabolic template was used to position formwork for the concrete and also to finish the surface (courtesy: CRAIA)

Fig. 2.24 The upgraded hole-in-the-ground antenna could survey consecutive strips of the sky by altering the position of the aerial mast. Here Gordon Stanley uses a theodolite to measure the mast's angle of tilt (courtesy: CRAIA)

tubes spaced at intervals of 1.8 m which were set in the concrete and constrained by fencing wire to the correct shape. As noted by Arthur Higgs, a senior member of the RP staff: "The actual reflecting surface … tolerance is within ±½ inch, so that the aerial is usable at all wavelengths down to 10 centimetres" (Higgs, 1953). Meanwhile, Stanley designed a new receiver, parts of which were housed in water-proof boxes at the base of the mast. The refurbished antenna was used to study 400 MHz emission from along the plane of the Galaxy and to survey for new radio sources. By this time, Dick McGee had become a vital member of the research team, as Bolton, Stanley and Slee also had to spend much of their time on the 12-Yagi array sky survey.

Upon thinking back on his Dover Heights experience, Bruce Slee (1978) felt that in general the equipment that he and other RP staff built was pretty reliable, but "After the War we had a lot of coal shortages and strikes – [so] for several years there we often had blackouts." Meanwhile, Gordon Stanley (1986) had more roman-tic views of those halcyon days in the late 1940s and early 1950s:

> In later years, we [he and John Bolton] regretted the 'loss of innocence' in the approach to research, compared to the business-like attitude that pervades radio astronomy today. Despite our later success at Owens Valley and Parkes, we never again experienced the exhilaration of those early years.

The remarkable hole-in-the-ground antenna was the last radio telescope built at Dover Heights, and from 1954 the field station was used briefly by the RP Cloud Physics group before being handed back to the Commonwealth Government in 1956. By this time, the Division's radio astronomical focus had shifted to the Potts Hill field station, and a new field station at Fleurs (Sections 2.6 and 2.10).

For further details on the role of the Dover Heights field station see Orchiston and Robertson (2017). For their personal reminiscences on the Dover Heights years see Bolton (1982), Slee (1994), Stanley (1994) and Westfold (1994).

2.3 Georges Heights

The Georges Heights radar station occupied an attractive strategic position on Middle Head, overlooking the entrance to Sydney Harbour (see Fig. 2.6), and dur-ing the war was home to a number of different radar antennas (Fig. 2.25). In 1947 and 1948, this site was used by RP as a short-lived field station (Orchiston, 2004a; Orchiston and Wendt, 2017; Wendt and Orchiston, 2018).

One of the wartime radar antennas at Georges Heights was an experimental unit featuring a 4.3 × 4.8 m section of a parabola and a cumbersome alt-azimuth mounting (Fig. 2.26), and this was used for early solar radio astronomy. The only way it could be used effectively was to place the antenna ahead of the Sun, let the Sun drift through the beam, hand-crank the antenna ahead of the Sun again, and repeat the process throughout the day. This procedure produced a distinctive 'picket fence' chart record. At this time, Payne-Scott and her collaborators were monitoring solar activity from

Fig. 2.25 The Georges Heights radar station and the view towards the entrance to Sydney Harbour (courtesy: CRAIA)

Fig. 2.26 Joe Pawsey (left) and Don Yabsley with the ex-radar antenna at Georges Heights. Yabsley and Fred Lehany used the antenna for solar observations at 600 and 1200 MHz (courtesy: CRAIA)

Fig. 2.27 One of the American 3.05 m AN/TPS-3 radar dishes undergoing testing at Georges Heights in August 1948. Two of these simple radio telescopes were transported to remote sites in Victoria and Tasmania to observe solar eclipses in 1948 and 1949 (courtesy: CRAIA)

Dover Heights, using 60, 100 and 200 MHz Yagis, and the Georges Heights antenna allowed the frequency coverage to be extended from 200 MHz to 600 and 1200 MHz.

Assigned to the Georges Heights antenna were two young RP radio engineers, Fred Lehany (1915–1980) and Don Yabsley (1923–2003). Lehany (1978) related that his involvement with this short-lived project "… came about in a typical 'Pawseyian way', before I knew what was happening … there was an observing program and … Yabsley and I were a suitable pair to share not only the week days but also the weekend duty …". This was Lehany's sole foray into radio astronomy, and soon afterwards he transferred to another CSIRO Division. In contrast, this proved the perfect radio astronomy training ground for a youthful Don Yabsley who later spent a decade working in the RP air navigation group, but returned to radio astronomy with the commissioning of the Parkes Radio Telescope. For a few months in 1947, Lehany and Yabsley were assisted by Bruce Slee, before he was re-assigned to Dover Heights.

In mid-1948, the Georges Heights field station was used as a test-base for two portable American ex-WWII 3.05 m dishes (Wendt and Orchiston, 2018) that were assembled in order to observe the 1 November 1948 partial solar eclipse from Rockbank (Victoria) and Strahan (Tasmania) (Fig. 2.27). Ironically, this eclipse was the death-knell for Georges Heights as a radio astronomy facility, for a decision was made to transfer the ex-radar antenna to the Potts Hill field station, where it could be used to monitor the eclipse. After less than two years as a field station, Georges Heights was closed down.

2.4 Hornsby Valley

Few if any residents now living in Hornsby would be aware that Hornsby Valley – Old Man Valley, as locals call it – was at one time an important astronomical site. Located on the northern outskirts of Sydney (Fig. 2.6) in a closed valley on farmland to the west of the Pacific Highway and near a quarry site, between 1947 and 1952 this RP field station was home to a number of unusual radio telescopes. Under the auspices of Frank Kerr (1918–2000), Ruby Payne-Scott, Alex Shain

---

Biobox 2.3: Alex Shain

Charles Alexander Shain (Fig. 2.28) was born in Sandringham, Victoria, on 6 February 1922. He completed a BSc (Second Class Honours) at the University of Melbourne in 1942. Following a brief stint of military service, Alex joined the Radiophysics Lab in November 1943 and worked on radar.

After the War, Shain was the leading 'low frequency' practitioner in the radio astronomy group and, assisted by Charlie Higgins, carried out Galactic observations at 9.15 MHz and 18.3 MHz from the Hornsby Valley field station. In 1956 Alex was responsible for the construction of the 19.7 MHz Shain Cross radio telescope at the Fleurs field station and used this to survey the Galactic Plane and to derive isophote plots of the strong discrete sources Centaurus A and Fornax A. Shain "… pioneered the study of two features peculiar to this range of radio astronomy: absorption in ionised interstellar hydrogen (HII regions), and absorption and refraction in the ionosphere" (Pawsey, 1960b: 244). Following the 1955 report by the Americans Bernard Burke (1928–2018) and Kenneth Franklin (1923–2007) of decametric emission from Jupiter, Alex revisited the 18.3 MHz 'static' recorded at Hornsby Valley in 1950–51, and to his chagrin discovered that some of these events were indeed Jupiter bursts! This was surely a missed research opportunity for Australian radio astronomy. When Alex Shain died from cancer on 11 February 1960, aged only 38, a promising career was cut short.

Fig. 2.28 Alex Shain (1922–1960) pioneered very low frequency radio astronomy, first at Hornsby Valley and then at the Fleurs field station (courtesy: CRAIA)

Fig. 2.29 The Hornsby Valley field station showing the antennas, transmission lines and instrument huts used by Kerr, Shain and Higgins in the 1948 Moon-bounce experiment. Radar signals at 17.84 and 21.54 MHz were transmitted from Shepparton (Victoria), bounced off the Moon, and received at Hornsby Valley (courtesy: CRAIA)

(1922–1960) and Charlie Higgins these were used to carry out pioneering studies in lunar, solar and Galactic astronomy (Orchiston and Slee, 2005; Orchiston et al., 2015a, 2015b).

In 1947 Frank Kerr and Alex Shain (Biobox 2.3) decided to study the structure of the upper part of the Earth's ionosphere with a simple rhombic antenna and modified communications receiver at the Hornsby Valley field station (Fig. 2.29). They succeeded in receiving radio signals broadcast at 17.84 and 21.54 MHz by Radio Australia (at Shepparton, Victoria) that had bounced off the Moon. This project lasted about one year, and was to be Kerr's sole foray into radar astronomy as he soon transferred to Potts Hill where he went on to make a name for himself through his hydrogen-line work.

Ruby Payne-Scott was keen to expand the solar work she had begun at Dover Heights, and towards the end of 1947 she moved to the Hornsby Valley field station and set up simple Yagi antennas for observations at 60, 65 and 85 MHz, together with an 18.3 MHz broadside array. She also made use of Kerr's 'Moon-bounce' rhombic antenna, operating at 19.8 MHz. Circular polarisation was investigated with the 85 MHz antennas, which consisted of a pair of crossed Yagis. This study ran from January through to September 1948, when Payne-Scott transferred to the Potts Hill field station (Goss and McGee, 2009).

After Kerr's departure to Potts Hill, Shain stayed on at Hornsby Valley, and he eventually became an acknowledged authority on low frequency radio astronomy, assisted throughout by Charlie Higgins. Up to this time, that region of the electromagnetic spectrum had been largely neglected because of the detrimental impact of the ionosphere on the incoming radio waves. Countering this problem was the simplicity of low frequency radio telescopes, where the ground itself could serve as a reflector. All that was required were posts to support the dipoles.

This simplicity was typical of the various arrays built at Hornsby Valley in 1949 and the early 1950s. For example, in 1949 Shain erected an array of eight half-wave dipoles strung between four rows of telegraph poles in order to investigate Galactic radiation at 18.3 MHz. The antenna system was attached to a standard communications receiver, and observations were carried out between May and November 1949.

The success of these early observations led Shain to expand the array to 30 horizontal half-wave dipoles, so that a more detailed survey could be carried out with a smaller beam. Although the antenna was stationary it was possible to move the beam electronically, and this meant that a wide strip of sky from declination −12° to −50° could be surveyed. The observations were carried out between June 1950 and June 1951, and only about 10% of all records were lost through interference or atmospherics.

In June 1951 the 18.3 MHz radio telescope was replaced by a network of 12 fixed horizontal half-wave dipoles, operating at 9.15 MHz. The network utilised some of the original telegraph poles, in four parallel banks of three antennas (Figs 2.30 and 2.31). The dipoles were attached to a standard communications receiver. Like its predecessor, this was a transit instrument which relied on the Earth's rotation in order to record radio emission from different strips of the sky (but this time without directing the beam). And once again, Sydney's fortuitous latitude meant that the celestial region of greatest interest, the Galactic Plane and centre of the Galaxy, would pass almost directly overhead. Between July 1951 and September 1952 this radio telescope was used by Shain to scan a strip of sky centred on declination −32°. Because of the prevalence of interference caused by the atmosphere and by radio stations, the most favourable time for observations was between midnight and half an hour before sunrise. It

Fig. 2.30 Some of the 9.15 MHz aerials at Hornsby Valley used in the low frequency sky survey centred on declination −32° (courtesy: CRAIA)

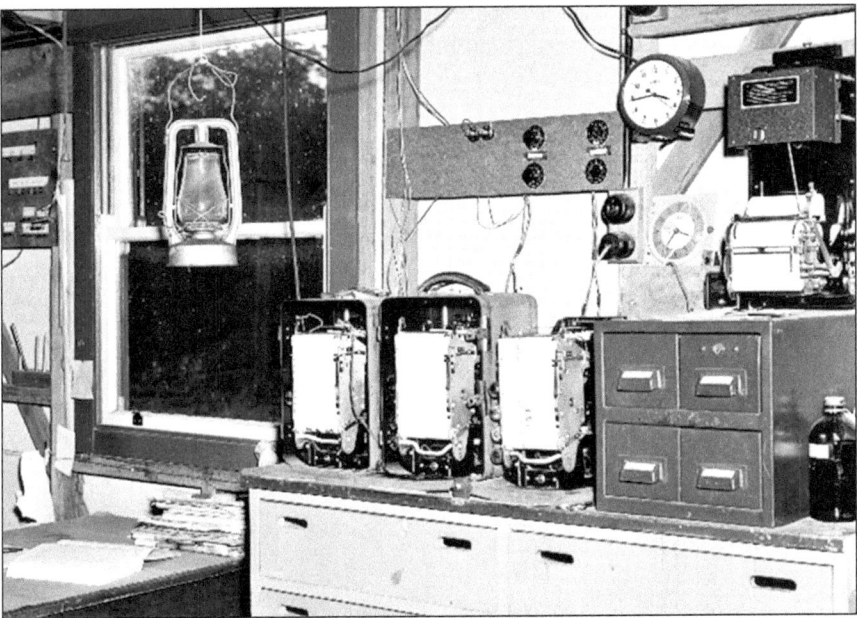

Fig. 2.31 Inside one of the receiver huts showing the chart recorders used for the 9.15 MHz sky survey in 1951–52 (courtesy: CRAIA)

Fig. 2.32 Radio astronomers attending the URSI General Assembly at the Hornsby Valley field station, in discussion with Alex Shain (far left) (courtesy: CRAIA)

was in August 1952, during this research project, that some of the delegates from the International Union of Radio Science Congress, which was meeting in Sydney, visited Hornsby Valley where they were given a guided tour of the site by Shain (Fig. 2.32).

As early as February 1952, Shain was having reservations about the future of the Hornsby Valley field station. The 9.15 MHz survey was underway, but interferometric observations planned at 18.3 MHz had been terminated prematurely when the equipment used was destroyed in a bush fire. After completion of the 9.15 MHz survey, Shain (1952) planned to embark on a major new low frequency project, but he concluded that "The Hornsby station will not be satisfactory for [this due to] ... lack of adequate open level space ... Hence we should plan to progressively evacuate Hornsby – breaking down the gear now in use after October 1952." While Shain suggested Badgery's Creek as a suitable site for the new low frequency work, it was in fact Fleurs that benefited from the eventual closure of the Hornsby Valley field station.

2.5   Bankstown Aerodrome

This was the last of the post-WWII RP field stations to be documented, and its very existence was only identified during doctoral research by Harry Wendt (2008). For about one year, during 1947–1948 Radiophysics maintained a field station at Bankstown Aerodrome, about 20 km south–west of the Sydney central business district. The aerodrome was set up in 1940 as a Royal Australian Air Force base, and after WWII some of the 18 hangars and 16 huts there "... were hired out to a range of enterprises, including the CSIR. This is how RP's little-known Bankstown Aerodrome field station was formed ..." (Wendt and Orchiston, 2019: 267–268).

By June 1947, RP had established a blossoming line of research in both solar (Orchiston et al., 2006) and cosmic noise investigations (Robertson et al., 2014) using sea-interferometry, but this technique could not locate the positions of short-duration sources, such as those associated with solar bursts. It was the task of Ross Fredrind Treharne (1919–1982; Treharne, 1983: 153), assisted by Alec Little (1925–1985; Mills, 1985), to develop a new type of interferometer that could detect these short-lived solar bursts.

By 20 February 1948 Treharne reported that he and Little had built a prototype 100 MHz two-element interferometer, and would soon start making solar observations, but shortly after the field station was vandalised and some of the equipment stolen. RP then decided to transfer the new interferometer to the secure and nearby much larger Potts Hill field station and the small Bankstown Aerodrome facility was closed down in July 1948. Treharne ended his brief flirtation with radio astronomy and went on to build a successful career in military research. The final development of what became known as the swept-lobe interferometer was then transferred to Ruby Payne-Scott and Little. As we will see in the next section, this unique radio telescope was used very effectively for daytime observations of the Sun, and of an evening by Bernie Mills to investigate discrete radio sources.

2.6   Potts Hill

The Potts Hill field station was located 17 km west–southwest of central Sydney (Fig. 2.6), at the radio-quiet site of the city's main water storage facility (Fig. 2.33). Begun in 1948 as a centre for solar radio astronomy, its role soon expanded to include important developments in non-solar radio astronomy (see Fig. 2.34 for site map). Radio astronomers associated at one time or another with Potts Hill included Chris Christiansen, Rod Davies (1930–2015), Jim Hindman, Frank Kerr, Alec Little, Don Mathewson (b. 1929), Bernie Mills (1920–2011), Harry Minnett, John Murray (b. ca. 1927), Ruby Payne-Scott, Jack Piddington, and a youthful Brian Robinson (1930–2004). The earliest radio telescopes on this site date from 1948, including a simple equatorially-mounted Yagi antenna used by Alec Little (Fig. 2.35) to monitor the Sun at 62 MHz. Another was a 3.05 m diameter dish (Fig. 2.27) with which Piddington and Minnett surveyed radiation from the region of the Galactic Centre at 1210 MHz (with a beam width of 2.9° at this frequency).

In the second half of 1948 the 4.3 × 4.8 m radar antenna at Georges Heights was relocated to Potts Hill (Orchiston and Wendt, 2017) and installed on an equatorial

Fig. 2.33 Aerial view of the eastern reservoir at Potts Hill from the north, site of one of RP's most productive field stations. Part of the larger, western water reservoir can be seen in the distance. The two 1420 MHz solar grating arrays are shown, with the original 32-element array in the background and the later 16-element array in the foreground (courtesy: CRAIA)

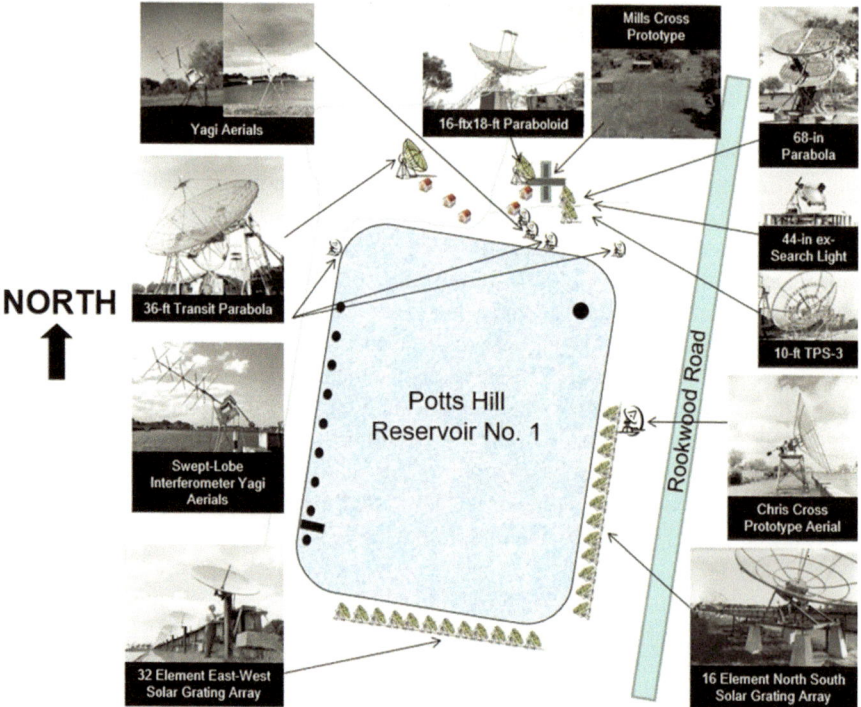

Fig. 2.34 Layout of the Potts Hill site showing the variety of radio telescopes in operation in the early 1950s (courtesy: Harry Wendt)

mounting (Fig. 2.36), after this was designated the station responsible for Sydney-based observations of the partial solar eclipse of 1 November. These observations were carried out at 600 MHz by Christiansen (Biobox 2.4), Yabsley and Mills. Other observations of this eclipse were made at Potts Hill by Piddington and Hindman at 3000 MHz, using an equatorially-mounted 1.7 m dish, which later was used with a distinctive full-aperture polarisation screen (Fig. 2.38). In addition, portable 3.05 m dishes operating at 600 MHz (see Fig. 2.27) were assembled and tested at Georges Heights and then transported to Rockbank, near Melbourne, and Strahan, on the west coast of Tasmania, to be used in conjunction with the ex-Georges Heights antenna at Potts Hill (Wendt and Orchiston, 2018). Following the celebrated New Zealand 'radio stars' field trip of June–August 1948, this eclipse therefore continued a tradition of establishing radio telescopes at temporary remote sites for special projects.

Fig. 2.35 At the Potts Hill
field station in 1949 (from
left): Ruby Payne-Scott,
Alec Little and Chris
Christiansen
(courtesy: CRAIA)

Fig. 2.36 After it was
relocated to Potts Hill in
1948, the ex-Georges
Heights radar antenna was
installed on an equatorial
mounting and used for a
wide range of research
projects. In 1951 it was the
first Australian radio
telescope to observe the
newly-discovered
hydrogen line
(courtesy: CRAIA)

Biobox 2.4: 'Chris' Christiansen

Wilbur Norman ('Chris') Christiansen (Fig. 2.37) was born in Melbourne on 9 August 1913 and died in Sydney on 26 April 2007. He studied mathematics and physics at the University of Melbourne, graduating with BSc and MSc (First Class Honours) degrees in 1934 and 1935 respectively. In 1953 the University awarded him a DSc, and he subsequently received an honorary DSc (Engineering) from the University of Sydney in 1980 and an honorary DEng from his *alma mater* in 1982. Immediately after completing his Masters, Chris worked for the Commonwealth X-ray and Radium Laboratory, before joining Amalgamated Wireless (Australasia) Ltd in 1937.

In 1948 Chris joined the Radiophysics Lab, rising to Senior Principal Research Scientist, before leaving in 1960 to accept a chair and head the Department of Electrical Engineering at the University of Sydney. After retiring in 1978, he was appointed a Visiting Fellow at Mount Stromlo and Siding Spring Observatories in Canberra. One of the pioneers of radio astronomy, Chris invented the solar grating interferometer at Potts Hill and the crossed grating interferometer at Fleurs, and also developed the innovative Fleurs Synthesis Telescope (Fig. 1.25). He was the first to use Earth rotation synthesis to produce a map of a celestial object, and provided confirmation of the existence of the 21 cm hydrogen line. He also carried out design work on radio telescopes in a number of overseas countries. A former President of URSI and Vice-President of the International Astronomical Union, Chris was a Fellow of the Australian Academy of Science, and was a recipient of the Syme Medal (University of Melbourne), the Russell Medal (Institution of Engineers Australia), the Fleming Award (Institution of Electrical Engineers) and the Adion Medal (Observatoire de Nice).

Apart from his many research papers, Chris was known for the textbook *Radiotelescopes* (CUP, 1969), co-authored by Jan Högbom (b. 1929). "Physicist, engineer, astronomer, teacher, administrator … Christiansen has in his times played many parts, and all of them with distinction" (Labrum, 1976; cf. data in Christiansen, 1976). For more on Christiansen's career see Swarup (2008), Davies (2009), Orchiston and Mathewson (2009), Wendt et al. (2011c), Frater and Goss (2011) and Frater et al. (2017). For his contributions to Indian and Chinese radio astronomy see Orchiston and Phakatkar (2019) and Shouguan (2017).

Fig. 2.37 Chris Christiansen (1912–2007) developed two innovative solar grating arrays at the Potts Hill field station and the Chris Cross at Fleurs (courtesy: CRAIA)

Fig. 2.38 The 1.7 m equatorially-mounted antenna was used in the late 1940s and early 1950s for solar monitoring at 3000 MHz, shown here with preparations underway for the November 1948 solar eclipse (courtesy: CRAIA)

This tradition was reinforced further in October 1949 for another solar eclipse (Wendt et al., 2008a). The portable dishes were located at Eaglehawk Neck on the east coast of Tasmania and near Sale in eastern Victoria, and employed in conjunction with the two Potts Hill antennas. In addition, the 1.1 m equatorially-mounted dish (Fig. 2.9) that had been used at the RP Lab for lunar and solar work in 1948–49 was available.

Not content with merely recording solar bursts, Ruby Payne-Scott and Alec Little wanted to record their positions, angular sizes and polarisation, so in early 1949 the RP workshop constructed a new interferometer consisting of three 97 MHz Yagi aerials (Fig. 2.39), aligned E–W, more or less along the northern edge of the eastern reservoir (Little and Payne-Scott, 1951). These were on equatorial mounts, which meant they could track the Sun for four hours each day centred on midday. Crossed dipoles allowed them to receive left-hand and right-hand circular polarisation. The signals were recorded as waves of various shapes on cathode ray tubes, where they were photographed. The Yagis could be used as either swept-lobe or fixed-lobe interferometers, depending on the type of investigation desired. Frank Kerr (1953) described how this interferometer "… was the first one in the world which could locate a source-position on the Sun sufficiently rapidly to be able to operate on the short-lived bursts." Meanwhile, of an evening this interferometer was used by Bernie Mills to investigate the positions of selected discrete sources.

Fig. 2.39 Joe Pawsey with one of the 97 MHz Yagi antennas used with the three-element position interferometer. The crossed dipoles allowed the polarisation of the solar bursts to be determined (courtesy: *Life* magazine)

According to Pawsey (1954), during the early 1950s solar monitoring continued at Potts Hill using 62, 97 and 200 MHz Yagi antennas (see Fig. 2.34), at 600 and 1210 MHz with the ex-radar antenna, and at 3000 and 9400 MHz using two small parabolas (e.g. see Fig. 2.38). The 62, 97 and 3000 MHz observations were terminated in 1956 when the radio astronomers involved became committed to other more urgent and time-consuming projects. The plan was to continue monitoring at the other frequencies until such observations were no longer required by staff at the Dapto and Fleurs field stations (Future Program for Radio Astronomy, 1958).

In 1949 non-solar research at Potts Hill expanded when Piddington and Minnett began an all-sky survey at 1210 MHz using the 3.05 m dish, and the following year they transferred the project to the much larger ex-radar antenna with a beam of ~1.4°, and also began observing at 3000 MHz using the 1.7 m dish, which – for this purpose – had been expanded to 2.3 m. At this frequency the dish had a beam width of 1.7°.

Solar astronomy at Potts Hill took a major step forward in 1951 when a 1420 MHz solar grating interferometer was constructed along the southern edge of the eastern water reservoir (Fig. 2.40). This innovative instrument was developed by Chris Christiansen, and consisted of 32 solid metal parabolic dishes each 1.8 m in diameter and spaced at 7 m intervals (Christiansen and Warburton, 1953). Each dish was on a simple equatorial mounting, but there was no drive. Instead, the operator had to manually change the position of each antenna during observations; to quote Christiansen (1976): "… as you run [continuously] from one end to the other you just click them so that you're following the Sun … [we] took it in turns. We'd do half an hour at a stretch … it was fairly strenuous ...". This novel radio telescope provided a series of 3 arcmin fan beams each separated by 1.7°. Since the Sun's

Fig. 2.40 Chris
Christiansen (above) with
the first Potts Hill solar
grating array, constructed
in 1951. There were 32
solid metal parabolic
dishes in the array, all
1.8 m in diameter. Each
dish was mounted on a
simple polar axis and could
be moved manually to
follow the Sun. During
observations the dish
positions were changed
approximately every
15 minutes. This involved
someone running down the
length of the array and
adjusting each of the 32
elements by hand. Later,
the entire array was
donated to a radio
astronomy group in India
(courtesy: CRAIA)

diameter is 30 arcmin, this meant that the Sun could only be in one beam at any one
time. The array was operational from February 1952, and was used daily for about
two hours, centred on midday to produce E–W scans of the Sun. Christiansen was
assisted throughout by Joe Warburton (1923–2005).

A second solar grating array was erected along the eastern margin of the same
reservoir in 1953 (Fig. 2.41). Although this also operated at 1420 MHz, it contained
just 16 equatorially-mounted mesh dishes 3.4 m in diameter. From September 1953
to April 1954 this was used to generate a series of N–S scans of the Sun in order to
investigate the distribution of radio emission from the 'quiet Sun'.

In 1954 two Indian visitors to Radiophysics, Govind Swarup (1929–2020; 2006)
and R. Parthasarathy, modified the E–W solar array so it could observe at 500 MHz,

Fig. 2.41 The 16-element north–south arm of the solar grating array in July 1953, located on the eastern bank of the Potts Hill reservoir. The E–W arm can be seen in the background on the southern bank (courtesy: CRAIA)

with fan beams of 8.25 arcmin each separated by 4.9°. This instrument was used between July 1954 and March 1955 to investigate emission from the quiet Sun at this new frequency. Soon after this project was completed, the 'Chris Cross' was constructed at Fleurs, and the Potts Hill solar program transferred to that field station. Arrangements were made to transfer ownership of the redundant Potts Hill grating array and the 500 MHz receiver to the National Physical Laboratory in India (Orchiston and Phakatkar, 2019). At this time the second solar grating array was also closed down.

One of the most interesting radio telescopes at Potts Hill was a prototype Mills Cross (Fig. 2.42) that was completed in early 1953 to test the cross-type telescope concept (Mills, 1953). Mills recalled that he "... had to convince people it would work, and there were also a number of basic problems I wasn't quite clear about myself which I wanted to experiment with ...". Despite his initial skepticism, RP Chief Edward G. ('Taffy') Bowen (1911–1991; 1973) was later to applaud Mills' initiative, describing how he "... had a brainwave ... literally a brainwave, and invented the Mills Cross ...". The one-tenth scale model built at Potts Hill consisted of N–S and E–W arms, each 37 m in length and containing 24 half-wavelength E–W aligned dipoles backed by a wire mesh reflecting screen. This novel instrument operated at 97 MHz and had an 8° beam, which could be swung in declination by changing the phases of the dipoles in the N–S arm (Mills and Little, 1953). Although some colleagues were skeptical, this prototype worked perfectly well – revealing the Galactic Centre and both Magellanic Clouds – and on this basis a new full-scale Mills Cross was erected at the new Fleurs field station in 1954.

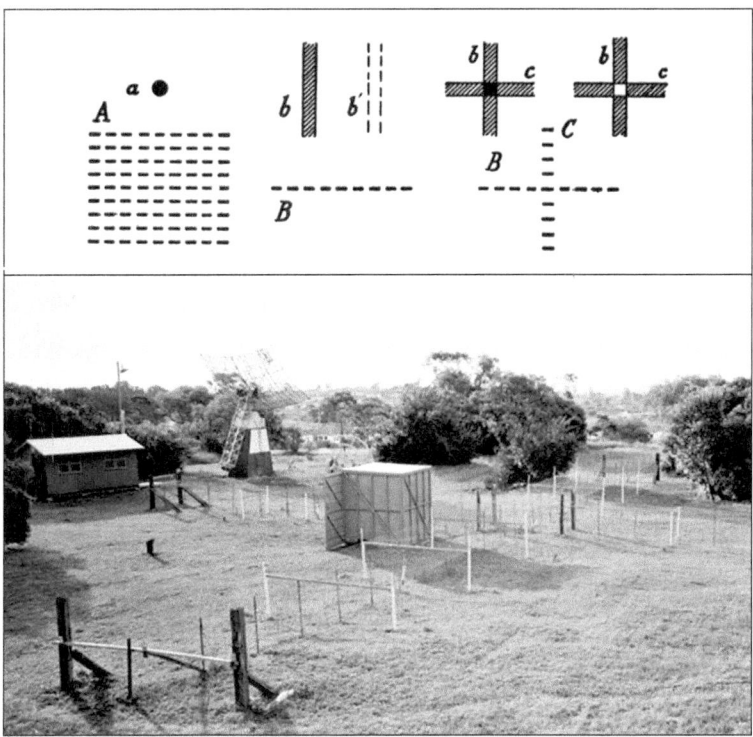

Fig. 2.42 (above) Diagram illustrating the cross-type radio telescope concept. (below) A proto-type Mills Cross was constructed quickly and cheaply at Potts Hill in order to test the concept – which it did with considerable success. This innovative new instrument had N–S and E–W arms 37 m long, and operated at a frequency of 97 MHz. The former Georges Heights radar antenna is in the background (courtesy: CRAIA)

With the emerging international interest in hydrogen-line work, a new, more suitable, radio telescope was required to replace the aging ex-Georges Heights radar antenna, and this came in the form of an 11 m dish constructed in 1952–53 (Fig. 2.43). The new dish had a beam width of 1.5°, and as a transit instrument could only be tilted in altitude. It was connected to a four channel receiver, each channel with a bandwidth of 40 kHz (Fig. 2.44). This facility was used extensively by Kerr, Hindman and Robinson for a range of Galactic and extragalactic hydrogen-line projects. In addition, this dish was used by Piddington and Trent for another project. In 1954–55 they carried out a survey of the sky at 600 MHz, using the former Piddington and Minnett 1210 MHz receiver, modified to operate at the lower frequency (the beam width was 3.3° at 600 MHz).

After making important contributions to solar, Galactic and extragalactic radio astronomy, the Potts Hill field station closed in 1963, when the remaining Galactic and extragalactic programs were transferred to Parkes. For detailed histories of the Potts Hill field station see Wendt (2008) and Wendt et al. (2008b, 2008c, 2011a).

Fig. 2.43 An 11 m dish was constructed at Potts Hill in 1952–53 and dedicated to research on the newly-discovered 1420 MHz hydrogen line. This dish replaced the aging ex-Georges Heights antenna (courtesy: CRAIA)

For personal reminiscences of their time at Potts Hill see Davies (2005) and Swarup (2008).

2.7 Penrith

In 1948 the need arose for a site where the spectra of solar bursts could be investigated using a novel type of radio telescope, and the short-lived Penrith field station was founded. This was located on a farm, close to the railway station at Penrith, a township to the west of Sydney at the foot of the Blue Mountains (see Fig. 2.6).

Fig. 2.44 The 4-channel
hydrogen-line receiver.
Taking readings is Brian
Robinson who joined the
H-line team in 1953
(courtesy: CRAIA)

Apart from a collection of simple huts which housed the electronic equipment, this field station featured just a single radio telescope, a large rhombic aerial.

This distinctive antenna (Fig. 2.45) consisted of a wooden cross-shaped frame measuring 10.7 × 7.1 m which supported the wire that detected the radio emission. The cross was anchored at one end, and in order to follow the Sun was first set to the correct position using crude hour angle and declination scales and then moved every 20 minutes or so by making adjustments to a number of different guy ropes. Joe Pawsey was widely regarded as an 'aerial guru', and it was he who provided the initial idea of using a rhombic antenna. Unfortunately, he was overseas when it was constructed, and Wild much later recalled that "... he was rather distressed when he came back and saw it. He had ideas of building it out of bamboo – he used to love bamboo (it's so nice and solid) – and he was horrified to see this rather heavy thing that was dragged around by ropes" (Wild, 1978). Nonetheless, it worked! This one antenna received solar radio emission over the frequency range 70–130 MHz, and displayed it on a cathode ray tube where it was photographed (see Wild and McCready, 1950). Apparently, the idea of attaching the antenna to a swept-frequency receiver came from Taffy Bowen, who was familiar with their use in a radar context during WWII.

The first serious scientific observations at the Penrith field station were full-time solar monitoring in February 1949. Paul Wild (1978) remembers that sometimes a week or more would go by without any solar activity, and that "... this involved an awful lot of work, traveling backwards and forward [from home] and watching, for

Fig. 2.45 (left) The rhombic antenna built by Paul Wild and Lindsay McCready at the Penrith field station. During 1949 this antenna operated with a swept-frequency spectrograph over a frequency range 70–130 MHz to study solar bursts. (right) The simple indicators used to point the antenna at the Sun at the start of each day's observations (courtesy: CRAIA)

nothing to happen – or so it seemed at the time. And then on other occasions there was huge tremendous activity." By the end of June the potential of this type of antenna had been proven and the search was on for a more 'radio-quiet' site where further antennas could be set up. This led to the founding of the Dapto field station, and in 1951 the short-lived Penrith field station was closed down. It had served its purpose.

Paul Wild (1923–2008) was the driving force behind the founding of the Penrith field station, which initiated his involvement in solar radio astronomy, a field in which he would soon become the world's leading authority (Biobox 2.5). Receiver development was in the hands of radio engineer John Murray and technician Bill Rowe; Murray would later play a leading role in developing instrumentation for RP's 21 cm hydrogen line research.

Biobox 2.5: Paul Wild

John Paul Wild (Fig. 2.46) was born in Sheffield, England, on 17 May 1923 and died in Canberra on 10 May 2008. He graduated BA in mathematics and physics from Cambridge University in 1943, and with an MA in 1946. During the latter part of WWII he served as a radio officer on the battleship *HMS King George V* and Sydney was the base for a number of Pacific operations. In 1947 Paul returned to Sydney where he was appointed a Research Officer in the Radiophysics Lab.

Wild specialised in radio studies of the Sun and he led the group studying short-lived phenomena such as violent flares in the solar atmosphere. He was

(continued)

Biobox 2.5 (continued)

Fig. 2.46 During the early 1950s Paul Wild (1923–2008) became the world's leading authority on solar radio astronomy (courtesy: CRAIA)

largely responsible for the initial design concepts of the Penrith and Dapto radio-spectrographs, the Dapto position interferometer and the Culgoora Radioheliograph (see Chapter 5). He moved quickly through the CSIRO ranks and in 1961 he was appointed a Chief Research Scientist. From 1971 to 1978 he was Chief of the Division of Radiophysics, and from 1978 to 1985 was the Chairman of CSIRO. Over the years Paul published a succession of seminal research papers, plus major review papers in Kuiper's *The Sun* (1953), *Advances in Electronics and Electron Physics* (1955) and in the 1963 issue of *Annual Review of Astronomy and Astrophysics* (co-authored by RP colleagues Steve Smerd and Alan Weiss).

Paul was widely regarded as one of the world's foremost solar physicists, and he received numerous awards, including the Jansky Lectureship (US National Radio Astronomy Observatory, 1973), the Herschel Medal (Royal Astronomical Society, 1974), the Lyle Medal (Australian Academy of Science, 1975) and the Royal Medal of the Royal Society of London (1980). In 1970 he was elected a Fellow of the Royal Society. He served as the President of the IAU Radio Astronomy Commission (C40) (1967–70) and as the Foreign Secretary of the Australian Academy of Science (1973–77). In 1978 Paul was appointed Commander of the Order of the British Empire (CBE) and in 1986 was made a Companion of the Order of Australia (AC). For further details see Stewart et al. (2011b), Frater and Ekers (2012) and Frater et al. (2017).

As an interesting aside it should be mentioned that in late 1947, when the Penrith field station was being planned, Ruby Payne-Scott (1947a) prepared a report for the RP 'hierarchy' on the "Possibility of Constructing a Spectrohelioscope or Spectroheliograph Suitable for Investigating Optical Correlation with Solar Radio Observations", where she advocated the use of a Lyot-type monochromatic filter.

But in the end it was CSIRO's Division of Physics that ultimately constructed this instrument. Under the guidance of their Chief, Dr Ron Giovanelli (1915–1984), there was close co-operation between this Division's solar optical astronomers and the solar radio astronomers at Radiophysics.

For in-depth accounts of the Penrith field station see Stewart (2009) and Stewart et al. (2010).

2.8 Dapto

This field station was located on a dairy farm just north of the township of Dapto and was set up in 1951 following a systematic search by Paul Wild. A memorandum prepared for Joe Pawsey by Wild on 16 January 1951 lists the criteria used in this search:

> The technical requirements of the site are:
>> (a) Rectangle of fairly level ground about 80 yds. In N-S direction by 25 yds. E-W. Room for expansion desirable.
>> (b) Radio interference in the frequency range 30–250 Mc/s to be as low as possible.
>> (c) 240 V mains supply very desirable. Telephone desirable.
> The "economic" requirements are:
>> (d) The site to be as close to Sydney as possible.
>> (e) Readily accessible by road, and public transport if possible.
>> (f) Efficient staffing arrangements of site to be possible.
>> (g) Reasonably proof against vandalism. (Wild, 1951)

The chief problem was finding a suitable, accessible, radio quiet site (i.e. one that satisfied condition (b) above). In addition to contemplating various sites down the NSW south coast almost to the Victorian border, Wild also looked north of Sydney as far as Wyong, to the southwest of Sydney at Picton, and south of Sydney at Moss Vale, Nowra and Kangaroo Valley. He also considered the existing field stations at Penrith and Hornsby Valley, but found levels of interference at both were unacceptable.

After reviewing various options, Wild successfully recommended Dapto for the new field station. The site was about 95 km south of Sydney (see Fig. 2.6) and readily accessed by road and rail; it was close to the owner's house, a key factor for security purposes; an abundance of flat land was available; the site was fringed by mountains to the north, screening it from Sydney-based radio interference; mains power and telephone were available; and various staff offered to live nearby or commute regularly from Sydney or Wollongong.

With the passage of the years, Dapto would contribute significantly to the international development of solar radio astronomy and, apart from Paul Wild, others who would play a leading role at this field station were John Murray, Jim Roberts (b. 1927), Kevin Sheridan (1918–2010), Steve Smerd (1916–1978) and, in later years, Shigemasa Suzuki (1920–2012).

Initially, the radio telescopes at Dapto consisted of three different crossed-rhombic aerials (Wild, 1951) in a N–S line, which covered the frequency ranges

Fig. 2.47 (above) The Dapto field station showing the three rhombic antennas and the main receiver building. (below) The rhombic antennas were mainly made of wood. Their crossed rhombic configuration allowed polarisation measurements to be made of the solar bursts (courtesy: CRAIA)

40–75, 75–140 and 140–240 MHz (Fig. 2.47). Each aerial was supported by an equatorial mounting, which meant it could be motor-driven to track the Sun. Meanwhile, the crossed-configuration allowed different polarisation measurements

Fig. 2.48 John Murray at
Dapto in 1952 with part of
the receiving equipment of
the radiospectrographs
(courtesy: CRAIA)

to be taken. Development of the aerials and the supporting receivers took time, so the
three solar radiospectrographs, as they were called, only became operational in
August 1952. Inside the receiver hut the signals from the three aerials went to three
swept-frequency receivers (Fig. 2.48), and then to cathode ray tubes where they were
photographed with cine cameras. This ingenious system allowed a complete spectrum to be obtained every half-second. Kevin Sheridan (1978) recalled that initially
it was hard to keep the facility operating successfully: "… it was a very difficult
piece of equipment to keep going … it was really a mechanical monster of sorts …".

Paul Wild (1978) recollected that once the radiospectrographs were ready, "…
we had a dreadful first three or four months, because we were right in the middle of
sunspot minimum. Nothing happened at all." Between 1958 and 1963 the Sun was
more active, and the solar facility was expanded when four new rhombic antennas
were added, allowing the lowest frequency received to successively be reduced
from 40 to 25 MHz (in 1958), 15 MHz (in 1960) and finally 5 MHz (in 1961). Then
in 1963 a 10 m parabolic dish with a log-periodic feed was installed, and this
allowed the upper frequency limit to be extended from 210 to 2000 MHz. By this
time, the Dapto facility had proved its worth and was no longer unique, for similar
radiospectrographs had been installed at Mitaka, Tokyo, in 1952 (see Nakajima
et al., 2014) and in 1957 at field stations by radio astronomy groups at Harvard
University (Thompson, 2010: 28) and the University of Michigan.

In the same year (1957), three new rhombic aerials were added at Dapto, in order
to investigate the positions, sizes and polarisations of different solar burst sources
over the frequency range 40–70 MHz, and two further rhombic aerials were added
in 1959. Four of these aerials, termed the position interferometer, were used for the
position and size studies, and all were located along an E–W line (Wild and Sheridan,
1958). The two most distant antennas were separated by 1 km. Although also made

Fig. 2.49 Paul Wild with one of the simple position interferometer rhombic aerials. Because these aerials remained stationary, they did not require equatorial mounts (courtesy: CRAIA)

with wooden frames, all four rhombics were of simpler design than the radiospectrograph aerials and did not need equatorial mounts as they remained stationary throughout the observations (see Fig. 2.49). The fifth new antenna was designed specifically for the study of polarisation. It comprised a crossed-rhombic aerial of identical dimensions to the position interferometer aerials, but was mounted equatorially so that it could track the Sun during the day. The system was set up so that the operator could manually switch between the position interferometer and the polarimeter, as required. After passing through the receivers the signals were initially displayed on a 30 cm cathode ray tube and photographed with a 35 mm cine camera, but in 1959 this arrangement was altered and the results were combined and preserved as a facsimile record (see Fig. 2.50).

These solar radio telescopes were used to good effect until the development of the Culgoora Radioheliograph, and were a tribute largely to Paul Wild's ingenuity, enthusiasm and leadership (Stewart et al., 2011b). In 1965 the Dapto field station was closed, which also meant the unfortunate loss of a proud institution at Radiophysics: the Dapto parties (see Fig. 2.51). Some of the Dapto aerials were relocated to Culgoora, but most remained at Dapto and were inherited by the University of Wollongong when it took over the site.

For in-depth accounts of the Dapto field station see Stewart (2009) and Stewart et al. (2011a).

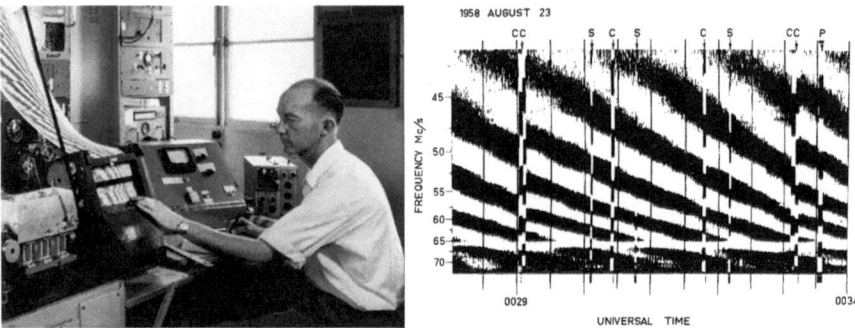

Fig. 2.50 Kevin Sheridan inside the receiver building operating the position interferometer facsimile machine. (right) a typical facsimile record showing a Type III solar burst (courtesy: CRAIA)

Fig. 2.51 The Dapto field station was famous for its parties. Leader of the Solar Group, Paul Wild is at front in the dark jersey, with Steve Smerd on guitar and 'Gib' Bogle on clarinet (courtesy: CRAIA)

2.9   Badgery's Creek

This field station was 50 km west–south–west of central Sydney (Fig. 2.6) on a CSIRO cattle research station, and was founded by Bernie Mills (Biobox 2.6) at the end of 1949 so that he could study discrete sources at 101 MHz, free from the electrical interference that had plagued him previously at Potts Hill.

Biobox 2.6: Bernie Mills

Bernard Yarnton Mills (Fig. 2.52) was born in Sydney on 8 August 1920 and died on 25 April 2011. He graduated with BSc, BE (Second Class Honours), ME (First Class Honours) and DScEng degrees from the University of Sydney in 1940, 1942, 1950 and 1959 respectively. In 1942 he joined the Radiophysics Lab and worked on wartime radar developments. After working briefly in other fields, he joined the radio astronomy group in 1948 and quickly showed a flair for developing innovative new aerial systems, such as the Badgery's Creek interferometer and the Mills Cross.

Along with Bruce Slee and Eric Hill, Bernie used the Mills Cross to discover over 600 discrete sources in the southern sky. At the same time the radio astronomy group in Cambridge published a sky survey claiming over 2000 new sources. However, there was very poor agreement between the two surveys, which led to a heated controversy between the Sydney and Cambridge groups as to which survey was in error. Eventually, it was shown that the great majority of the Cambridge sources were fictitious, the result of an instrumental effect known as 'confusion'.

Mills left Radiophysics in 1960 to accept a Readership at the University of Sydney and was later promoted to Professor of Physics (Astrophysics). Bernie was able to secure funding from the US National Science Foundation to construct a much larger version of the Mills Cross near Canberra (see Chapter 5). He received the Lyle Medal from the Australian Academy of Science in 1957, the Britannica Australia Award for Science in 1967 (shared with John Bolton), and the Grote Reber Medal for Radio Astronomy in 2006. He was elected a Fellow of the Australian Academy of Science in 1959 and four years later joined an elite band of Australian radio astronomers when he became a Fellow of the Royal Society. In 1976 Bernie received Australia's highest civil award when he was made a Companion of the Order of Australia (AC). For further details see Frater et al. (2013, 2017), Mills (2006) and Orchiston (2014).

Fig. 2.52 Bernie Mills (1920–2011) invented a new cross-type radio telescope, which inspired several radio astronomy groups in Europe and the United States to build their own versions of the Mills Cross (courtesy: CRAIA)

Fig. 2.53 The Badgery's
Creek field station featured
three 100 MHz broadside
arrays positioned along an
E–W line
(courtesy: CRAIA)

Initially there were three identical radio telescopes at this site: simple broadside arrays positioned along an E–W line, and mounted so they could be tilted about their E–W horizontal axes (Fig. 2.53). Each antenna contained 24 half-wave dipoles, backed by a reflecting screen. Antenna 2 was located about 60 m to the east of Antenna 1, with Antenna 3 a further 210 m to the east. All cables between the antennas and the receiver hut were buried in order to reduce fluctuations in electrical strength associated with temperature variations. Two different receivers were used with the antennas, so that the outputs of any two aerial spacings could be recorded simultaneously (Fig. 2.54). Between February and December 1950 Mills used this interferometer to conduct a survey of the Galactic distribution of discrete sources.

Mills then wanted to study the positions and angular sizes of four of the Galactic sources, and to do this he set up a two-element variable baseline interferometer. This

Fig. 2.54   Bernie Mills and the receiving equipment at Badgery's Creek (courtesy: CRAIA)

Fig. 2.55   The mobile two-element Yagi array and microwave link formed part of the variable baseline interferometer at Badgery's Creek. This array was used in conjunction with the broadside arrays to study four of the brightest discrete sources at 101 MHz (courtesy: CRAIA)

used one of the three broadside arrays, plus a mobile two-element Yagi array (operating at 101 MHz) and a microwave link (Fig. 2.55). This was the first time such a link had been used in radio astronomy, and to Mills (1976) was simply a matter of logic:

If we wanted to try different spacings and different places, then obviously you couldn't go coiling and uncoiling miles of cables. The radio link was the obvious way of doing it … and the only real thing I was worried about was getting this compensating delay which I realised was necessary.

During 1952 observations were made of four strong discrete sources at nine different E–W spacings (ranging from 0.3 to 10 km), and three spacings in other directions. At the end of this project Mills transferred to Potts Hill, where he would build the world's first prototype cross telescope. Much later, he explained that

This [Badgery's Creek] survey was actually the basis for the [Mills] Cross because I realised that it was necessary in any survey to have an instrument which would respond to close spacings and large angular size structure. Otherwise, one would simply miss it and miss a lot of the information available in the sky. And it was as a result of this survey that I thought of the Cross as being the sort of thing one must use. One must use pencil beams for surveys. That was the basic idea I had in mind. (Mills, 1976).

Meanwhile, the Badgery's Creek field station was retained by Radiophysics and subsequently used by some of the radio astronomers based at Fleurs, before it was finally closed in 1956.

2.10   Fleurs

Fleurs was situated about 40 km west–south–west of central Sydney (Fig. 2.6), and occupied an expanse of flat land between South Creek and Kemps Creek adjacent to a WWII air strip near Badgery's Creek. Between 1954 and 1963, Fleurs was one of RP's leading field stations as home to three innovative cross-type radio telescopes, the Mills Cross, the Shain Cross and the Chris Cross (Fig. 2.56). Leading radio astronomers associated with this site were Chris Christiansen, Eric Hill (1927–2016), Norman Labrum, Bernie Mills, Dick Mullaly (1924–2001), Alex Shain, Kevin Sheridan and Bruce Slee, assisted by others, including Charlie Higgins and Wayne Orchiston (b. 1943) (see Orchiston and Slee, 2002b).

The first radio telescope at Fleurs was the Mills Cross (Mills et al., 1958), which was constructed during 1953–54 (Figs 2.57 and 2.58) following the success of the small-scale prototype erected at Potts Hill in 1953. In July 1954 Pawsey identified this as RP's number one priority project, and although progress was excellent, "… all available manpower is required to help it along." The cost of constructing the aerial and receiver hut (excluding labour) was the not inconsiderable sum of £14,000 (Mills, 1953). The cross boasted 460 m long N–S and E–W arms, each containing two rows of 250 half-wave dipoles (see Fig. 2.59). The Mills Cross operated at a frequency of 85.5 MHz giving a 49 arcmin beam – which in those days was regarded as remarkable. Signals from the two arms were channeled to the central hut where the receivers and other equipment were located (Figs 2.60 and 2.61).

Although this radio telescope was effectively a transit instrument in that neither arm was fully steerable, by altering the phasing of the dipoles in the N–S arm it was possible to observe different declination strips of sky. Alec Little played a key role

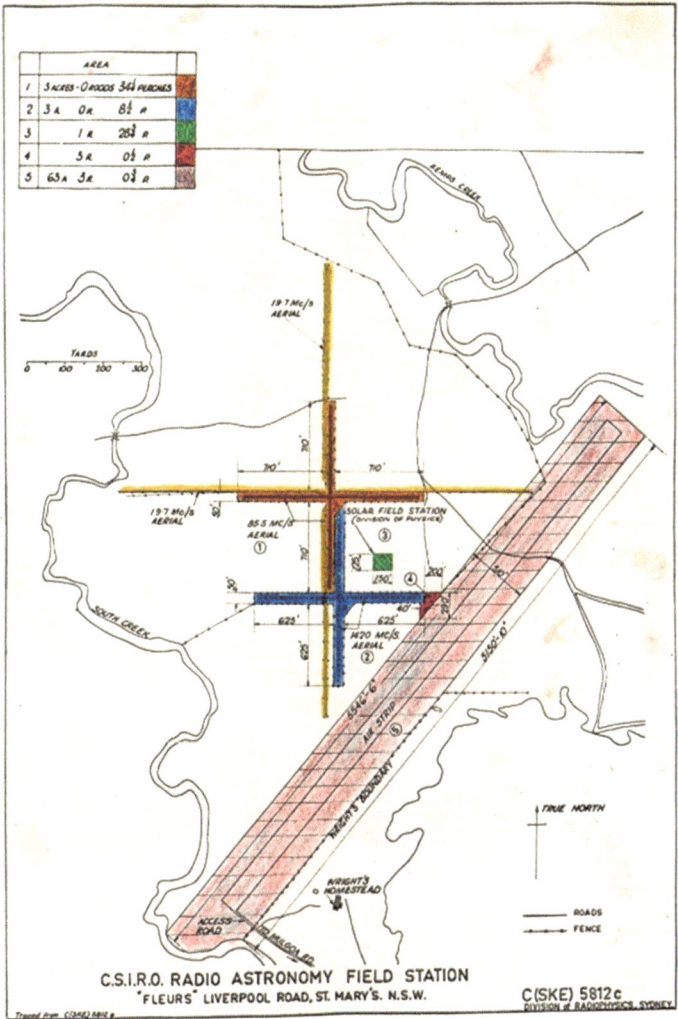

Fig. 2.56 Fleurs site plan showing the locations of the Mills (brown), Shain (yellow) and Chris (blue) Crosses in relation to the two creeks and the airstrip (courtesy: CRAIA)

throughout the design, construction and testing phases (see Fig. 2.62) and Kevin Sheridan assisted with the design and construction of the receiver. Later he reminisced: "I would say we spent about a year just getting it all together, and then two or three months polishing it off. It worked fairly well; it was a fairly good system …" (Sheridan, 1978). From 1954 to 1957 this "fairly good system" was used by Mills, Sheridan, Slee and Hill to conduct an all-sky survey of discrete sources, and to investigate radio emission from selected sources. Some of the results had important cosmological implications which were to involve Australia in a good deal of international controversy, as we shall see in Chapter 4.

Fig. 2.57 The Mills Cross under construction at Fleurs in 1953–54 (courtesy: CRAIA)

Completion of the Mills Cross sky survey did not spell the end for this radio telescope (see Slee, 2005). In 1958–59 Bruce Goddard, Arthur Watkinson and Bernie Mills used the E–W arm of this antenna, a 91 m section of the S arm of the Cross, and an identical 50-dipole N–S array at nearby Wallacia (Fig. 1.14) to carry out an 85.5 MHz investigation of the sizes of some of the smaller sources in the Mills–Slee–Hill catalogues. In addition, in June 1957 Bruce Slee used the N–S arm of the Cross, at 85.5 MHz, and two helical aerials located about 200 m to the east of the Cross to investigate scattering by solar coronal irregularities. Between June and October 1960, he extended this study using the E–W arm of the Mills Cross in conjunction with the array at Wallacia. From September 1960 to May 1962, Bruce Slee and Charlie Higgins used the N–S arm of the Mills Cross to carry out pioneering observations of radio emission from a number of nearby flare stars. Finally, during

Fig. 2.58 RP photographer Ken Nash uses 'Flo' as a vantage point to photograph the new Mills Cross at Fleurs. Ken took many of the photographs reproduced in this book (courtesy: CRAIA)

Fig. 2.59 Aerial view of the completed Mills Cross, looking south, taken in October 1954. The receiver hut is near the junction of the two arms (courtesy: CRAIA)

1961–62, Slee and visiting Cambridge radio astronomer, Peter Scheuer (ca.1930–2001), used the E–W arm of the Mills Cross and barley-sugar arrays erected temporarily at Cumberland Park, Rossmore, Llandilo and Freeman's Reach

Fig. 2.60 Central area of the Mills Cross showing the receiver hut and the microwave antenna used for communicating with remote sites involved in interferometry (courtesy: CRAIA)

Fig. 2.61 Inside the hut showing the Mills Cross receiving equipment in November 1955, with Bruce Slee examining one of the chart records (courtesy: CRAIA)

(respectively 6 and 10 km south, and 17 and 32 km north of Fleurs – see Fig. 1.14) to research the sizes of selected sources in the Mills–Slee–Hill catalogues.

During 1955 a 19.6 MHz two-element E–W interferometer was constructed at Fleurs. Each aerial consisted of four full-wave dipoles suspended between telegraph poles and separated by 183 m. The aerials were attached to two receivers, one with the aerials connected in phase, and the other out of phase. Observations could be

Fig. 2.62 Alec Little (left) and Bernie Mills taking phase and amplitude readings at different dipoles along the Mills Cross (courtesy: CRAIA)

made for about five hours each day. In addition, in October 1955 two simple radio telescopes operating at 14 and 27 MHz were constructed. Between June 1955 and March 1956, the 19.6 MHz E–W interferometer was used by Alex Shain and Frank Gardner to investigate burst emission from Jupiter (see the next chapter). Parallel observations with the 14 and 27 MHz radio telescopes were also made from November 1955 to March 1956.

In 1956, just two years after the Mills Cross became operational, a much longer low frequency antenna was completed nearby (see Fig. 2.63). This was the Shain Cross (see Shain, 1958), and as early as mid-1954, Pawsey had designated this as RP's number three priority project (after completion of the Fleurs Mills Cross and the Murraybank 48-channel H-line receiver). Given its distinctive name, there is no surprise that Alex Shain was largely responsible for this radio telescope, but Alec Little did help him with the antenna design and Kevin Sheridan with the receiving equipment. This new radio telescope, operating at a frequency of 19.7 MHz and with a beam width of 1.5°, was built alongside the Mills Cross. Its construction was spurred on by news received from Bernie Mills in mid-1954 that the Department of Terrestrial Magnetism in Washington, DC had just completed a 20 MHz Mills Cross of its own, with arms 634 m long, at its Maryland field station.

Although the Shain Cross had much longer arms than the Mills Cross, its construction was much simpler in that it comprised a series of dipoles strung 4 m above the ground between telegraph poles, with the ground serving as a reflector. The N–S arm was 1151 m in length and contained 151 dipoles, with the main supporting telegraph poles spaced at 46 m intervals; lighter poles at 23 m intervals provided further support and carried the feeder lines. The E–W arm was a little shorter at 1036 m,

Fig. 2.63 (left) Section of the 19.7 MHz Shain Cross at Fleurs, also showing part of the N–S arm of the Mills Cross and the broadside array used for long baseline interferometry. (right) Alex Shain making adjustments to one of the dipole phase switches located along the N–S arm of the Shain Cross. The switches allow the beam to be directed to different E–W positions (courtesy: CRAIA)

contained 132 dipoles, and featured a 30 m gap in the middle. This innovative new radio telescope evolved out of Shain's earlier exploits at the Hornsby Valley field station, and the pilot observations that he and Gardner had made at Fleurs. By exploiting the cross-type principle, Shain was able to use the new Cross to carry out a survey of the Galactic Plane and to monitor decametric burst emission from Jupiter.

After Shain's untimely death in 1960, Bruce Slee and Charlie Higgins continued to use Fleurs for low frequency research. Late in 1961 they used the N–S arm of the Shain Cross to search successfully for radio emission from flare stars. In August 1962 they erected a square array of 19.7 MHz dipoles at Fleurs and an identical array at Freeman's Reach, 32 km to the north in order to investigate the size of the region responsible for the Jovian bursts. In 1963–64 they expanded this project by setting up radio-linked arrays at Dapto and Jamberoo far to the south of Sydney, and at Heaton to the north (see Fig. 2.8).

Fleurs gained its third large radio telescope in 1957 – just in time for the International Geophysical Year – when a major new solar array, the Chris Cross (Christiansen et al., 1961), was completed at a cost of about £21,000 (Costs – Chris Cross, 1957). Named, appropriately, after 'Chris' Christiansen, this innovative instrument comprised 433 m N–S and E–W arms each containing 32 parabolic dishes 5.8 m in diameter (Fig. 2.64). These dishes were constructed of wire mesh suitable for operation at 1420 MHz (21 cm), and were equatorially-mounted to be able to track the Sun. The Chris Cross was constructed adjacent to the Mills and Shain Crosses (Fig. 2.65), and combined the principles of the Mills Cross and the solar grating interferometer. It was the first radio telescope in the world to generate daily two-dimensional high-resolution radio images of the Sun. For more on the Chris Cross see Orchiston (2004b) and Orchiston and Mathewson (2009).

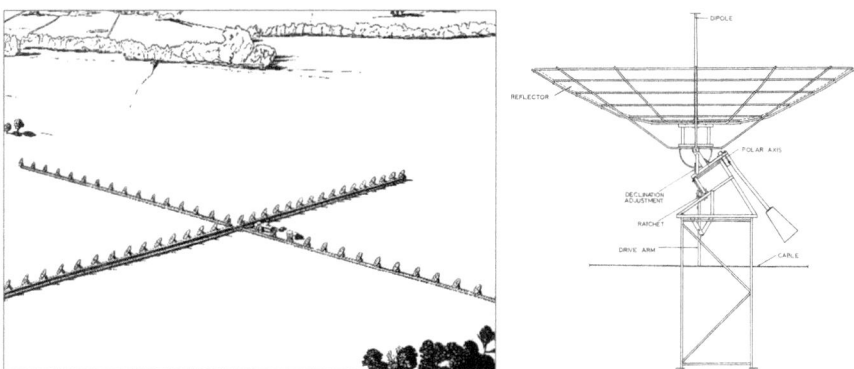

Fig. 2.64 Artist's impression of the Chris Cross and the design details of the individual antennas (courtesy: CRAIA)

Fig. 2.65 View looking south showing sections of all three cross telescopes at Fleurs (from left): antennas in the N–S arm of the Chris Cross, the white woodwork of the Mills Cross and the poles of the Shain Cross (courtesy: Sullivan collection)

A further development occurred at Fleurs in 1959 when an 18 m paraboloid, known as the 'Kennedy Dish' after its American manufacturer, was installed at the eastern end of the Chris Cross (Fig. 2.66). As early as 1956 Pawsey and other RP staff had been referring to the acquisition of "... a large aerial (e.g. 60 ft) ...", intended specifically for H-line work, but in fact this new dish ended up being used for continuum observations. When combined with the dishes of the E–W arm of the

Fig. 2.66 View looking west showing the 18 m Kennedy Dish, installed at Fleurs in 1959, and some of the antennas in the E–W arm of the Chris Cross (courtesy: CRAIA)

Chris Cross, the Kennedy Dish provided the Fleurs radio astronomers with a power-ful new compound interferometer, which boasted a 1.5 arcmin fan beam at 1420 MHz (Orchiston and Mathewson, 2009). Of an evening – when the Chris Cross was not operational – this array was used to investigate some of the strongest discrete sources.

A major change took place at Fleurs at the end of 1962 when the Kennedy Dish was transferred to Parkes (Orchiston, 2012) to be used in conjunction with the 64 m Parkes Radio Telescope (see Fig. 1.43). By this time the research programs that justified the construction of the Mills and Shain Crosses had come to an end and Fleurs had served its purpose. No longer required as a field station, on 1 July 1963 the lease was formally transferred to the University of Sydney. The School of Electrical Engineering under the direction of its newly appointed professor, Chris Christiansen, then spent the next decade converting the Chris Cross into the Fleurs Synthesis Telescope – but more on this in Chapter 5.

In summary, during the ten-year interval 1954–63, Fleurs was one of the world's foremost radio astronomy sites, and it played an important role in furthering solar, Galactic and extragalactic radio astronomy. RP Chief, Taffy Bowen (1984: 97) went so far as to claim that the three Fleurs cross-type radio telescopes "… were among the great successes of the 1950s and were responsible for a large part of the Division's research output over that period." They consolidated the international standing of Christiansen and Mills, helped build the emerging reputations of people like Shain, Sheridan and Slee, and served as stepping stones to the Division's next major advances in instrumentation: the Parkes Radio Telescope and the Culgoora

Fig. 2.67 Interior of the mobile field laboratory 'Flo' that was fitted out with a variety of instrumentation (courtesy: CRAIA)

Radioheliograph. Before leaving this account of Fleurs it is only proper that we acknowledge the invaluable contribution of 'Flo', the modest little mobile field laboratory (see Figs 2.3, 2.58 and 2.67) that serviced all three cross-type radio telescopes. Flo was a major resource throughout the lifetime of the field station, and for those of us who enjoyed our Fleurs experiences, life without Flo was simply unthinkable.

2.11 Murraybank

The Murraybank field station was located in the Sydney suburb of West Pennant Hills (see Fig. 2.6), on an orchard ('Rosebank') owned by the father of RP radio astronomer, John Murray. It was set up in 1956 to carry out H-line observations with a new purpose-built radio telescope and receiver. In a review of the Radio Astronomy Group research programs, Pawsey (1954) specifically mentioned the clear need for first class equipment if the H-line research is to prosper: "The current instrumental deficiency is a receiver giving an instantaneous profile." In October 1956, he reported that as other work tapered off, it would soon "… be possible to apply our facilities to a new aerial for the H-line work … [and] develop the new receiver …" (Pawsey, 1956). Murraybank was the proud recipient of this new equipment, and Dick McGee and John Murray were the two RP radio astronomers who spent most time at this field station.

Fig. 2.68 Installation of the 6.4 m parabolic antenna and adjacent receiver hut at the Murraybank field station. This site in West Pennant Hills was located on an orchard belonging to John Murray's father, and so the name Murraybank (courtesy: CRAIA)

Previous H-line work had taken place at Potts Hill, first with the ex-Georges Heights radar antenna and then with the 11 m transit dish, but what was needed was a new steerable dish dedicated to H-line work at 1420 MHz. At 6.4 m the Murraybank radio telescope (Fig. 2.68) was considerably smaller than its Potts Hill predecessor, but its alt-azimuth mounting meant that interesting areas of the sky could be accessed at will. At 1420 MHz, the beam width was 2.2°.

There was also a marked improvement in the receiving equipment, as the old H-line unit at Potts Hill was replaced by a new 48-channel receiver built mainly by Murray and McGee (Fig. 2.69). This contained 44 separate narrow-band channels, spread at 33 kHz intervals across the frequency 1420 MHz, and four wide-band channels at either end of the range, which were used to obtain reliable zero levels. In mid-1958 the intention was to eventually transfer this new receiver to Fleurs and the Kennedy Dish, but this never happened.

With this new receiver, a complete H-line profile could be obtained in just two minutes, sixty times faster than with the old Potts Hill equipment. Reduction of the large amounts of observational data produced posed a major challenge and relied on the development of early computers (Fig. 1.11). In the main, the Murraybank facility was used by McGee and Murray to investigate neutral hydrogen from the Large and Small Magellanic Clouds and in the Taurus–Orion region, and to conduct an all-sky H-line survey.

Fig. 2.69 Dick McGee
(left) and John Murray
examine a chart record at
Murraybank
(courtesy: CRAIA)

For the Radiophysics Lab, Murraybank served an important role as the test-bed for the innovative 48-channel H-line receiver, and this field station only closed down when the receiver was transferred to the newly-opened Parkes Radio Telescope at the end of 1961 (Brooks and Sinclair, 1994).

For detailed accounts of the Murraybank field station see Wendt (2008) and Wendt et al. (2011b).

References

Bolton, J.G., 1982. Radio astronomy at Dover Heights. *Proceedings of the Astronomical Society of Australia*, 4, 349–358.

Bolton, J.G., 1986. Letter to Woody Sullivan, dated 13 December. Sullivan collection.

Bowen, E.G., 1973. Taped interview with Woody Sullivan, dated 24 December.

Bowen, E.G., 1984. The origins of radio astronomy in Australia. In Sullivan, W.T., 1984. Pp. 85–111.

Briton, J., 1945. Letter to G. Gresford, dated 6 November, file A1/1/1 'Radio Astronomy – General Correspondence'.

Brooks, J., and Sinclair, M., 1994. Receivers, electronics and people – past and present. In Goddard, D.E., and Milne, D.K. (eds). *Parkes – Thirty Years of Radio Astronomy*. Melbourne, CSIRO Publishing. Pp. 47–56.

Christiansen, W.N., 1976. Taped interview with Woody Sullivan, dated 27 August.

Christiansen, W.N., 1984. The first decade of solar radio astronomy in Australia. In Sullivan, W.T., 1984. Pp. 113–131.

Christiansen, W.N., 1986. Undated letter to Woody Sullivan, written in June. Sullivan collection.

Christiansen, W.N., and Warburton, J.A., 1953. The distribution of radio brightness over the solar disk at a wavelength of 21 cm. Part 1. A new highly directional aerial system. *Australian Journal of Physics*, 6, 190–202.

Christiansen, W.N., Labrum, N.R., McAlister, K.R., and Mathewson, D.S., 1961. The cross-grating interferometer: a new high-resolution radio telescope. *Proceedings of the Institution of Electrical Engineers*, 108B, 48–55.

Costs – Chris Cross, 1957. Undated report. Sydney, RP. Sullivan collection.

Davies, R.D., 2005. A history of the Potts Hill radio astronomy field station. *Journal of Astronomical History and Heritage*, 8, 87–96.

Davies, R.D., 2009. Recollections of two and a half years with 'Chris' Christiansen. *Journal of Astronomical History and Heritage*, 12, 4–10.

Dillett, T., 2000. *The Royal Australian Air Force on Collaroy Plateau in the Second World War.* Collaroy (Sydney), J.E. & D.M. Dellit.

Frater, R.H., and Ekers, R.D., 2012. John Paul Wild AC CBE FAA FTSE. 17 May 1923 – 10 May 2008. *Biographical Memoirs of Fellows of the Royal Society*, 58, 327–346.

Frater, R.H., and Goss, M.W., 2011. Wilbur Norman Christiansen 1912–2007. *Historical Records of Australian Science*, 21, 215–228.

Frater, R.H., Goss, W.M., and Wendt, H.W., 2013. Biographical Memoir: Bernard Yarton Mills 1920 – 2011. *Historical Records of Australian Science*, 24, 294–315.

Frater, R.H., Goss, W.M., and Wendt, H.W., 2017. *Four Pillars of Radio Astronomy: Mills, Christiansen, Wild, Bracewell.* Cham (Switzerland), Springer.

Future Program for Radio Astronomy, 1958. Sydney, RP. Sullivan collection.

Gardner, F., 1973. Taped interview with Woody Sullivan, dated 21 February.

Gardner, F., 1986. Letter to Woody Sullivan, dated 12 November. Sullivan collection.

Goss, W.M., 2013. *Making Waves, The Story of Ruby Payne-Scott, Australian Pioneer Radio Astronomer.* Heidelberg, Springer.

Goss, W.M., and McGee, R.X., 2009. *Under the Radar – The First Woman in Radio Astronomy: Ruby Payne-Scott.* Heidelberg, Springer.

Higgs, A., 1953. Letter to A.R. Cohen, dated 30 October. RP Archives (A1/3/1)

Hindman, J.V., 1978. Taped interview with Woody Sullivan, dated 14 March.

Kellermann, K.I., 1996. John Gatenby Bolton (1922–1993). *Publications of the Astronomical Society of the Pacific*, 108, 729-737.

Kerr, F.J., 1953. *Sydney Water Board Journal*, 126–127.

Kerr, F.J., 1971. Taped interview with Woody Sullivan, dated 3 October.

Labrum, N.R., 1976. W.N. Christiansen. MS prepared in connection with the award of the Adion Medal. Sullivan collection.

Lehany, F.J., 1978. Taped interview with Woody Sullivan, dated 13 March.

Little, A.G., and Payne-Scott, R., 1951. The position and movement on the solar disk of sources of radiation at a frequency of 97 Mc/s. 1. Equipment. *Australian Journal of Scientific Research*, A4, 489–507.

Mills, B.Y., 1953. Large aerial for metre wavelengths. RP Report. Sullivan collection.

Mills, B.Y., 1976. Taped interview with Woody Sullivan, 25–26 August.

Mills, B.Y., 1985. Obituary – Little, Alec. *Proceedings of the Astronomical Society of Australia*, 6, 113.

Mills, B.Y., 2006. An engineer becomes an astronomer. *Annual Review of Astronomy and Astrophysics*, 44, 1–15.

Mills, B.Y., and Little, A.G., 1953. A high-resolution aerial system of a new type. *Australian Journal of Physics*, 6, 272–278.

Mills, B.Y., Little, A.G., Sheridan, K.V., and Slee, O.B., 1958. A high resolution radio telescope for use at 3.5 m. *Proceedings of the Institute of Radio Engineers*, 46, 67–84.

Nakajima, H., Ishiguro, M., Orchiston, W., Akabane, K., Enome, S., Hayashi, M., Kaifu, N., Nakamura, T., and Tsuchiya, A., 2014. Highlighting the history of Japanese radio astronomy. 3: Early solar radio research at the Tokyo Astronomical Observatory. *Journal of Astronomical History and Heritage*, 17, 2–28.

Nakamura, T., and Orchiston, W. (eds), 2017. *The Emergence of Astrophysics in Asia: Opening a New Window on the Universe.* Cham (Switzerland), Springer.

Orchiston, W., 1994. John Bolton, discrete sources, and the New Zealand field-trip of 1948. *Australian Journal of Physics*, 47, 541–547.

Orchiston, W., 2004a. Radio astronomy at the short-lived Georges Heights field station. *ATNF News*, 52, 8–9.

Orchiston, W., 2004b. The rise and fall of the Chris Cross: a pioneering Australian radio telescope. In Orchiston, W., et al. (eds). *Astronomical Archives and Instruments in the Asia–Pacific Region.* Seoul, Yonsei University. Pp. 157–162.

Orchiston, W., 2012. The Parkes 18-m antenna: a brief historical evaluation. *Journal of Astronomical History and Heritage*, 15, 96–99.

Orchiston, W., 2014. Mills, Bernard Yarnton. In Hockey, T. et al. (eds). *Biographical Encyclopedia of Astronomers. Second Edition.* New York, Springer. Pp. 1482–1484.

Orchiston, W., and Kellermann, K.I., 2008. Bolton, John Gatenby. In *Dictionary of Scientific Biography.* New York, Gale. Pp. 332–337.

Orchiston, W., and Mathewson, D., 2009. Chris Christiansen and the Chris Cross. *Journal of Astronomical History and Heritage*, 12, 11–32.

Orchiston, W., and Phakatkar, S., 2019. A tribute to Professor Govind Swarup, FRS: the father of Indian radio astronomy. *Journal of Astronomical History and Heritage*, 22, 3–44.

Orchiston, W., and Robertson, P., 2017. The origin and development of extragalactic radio astronomy: the role of the CSIRO's Division of Radiophysics Dover Heights Field Station in Sydney. *Journal of Astronomical History and Heritage*, 20, 289–312.

Orchiston, W., and Slee, B., 2002a. Ingenuity and initiative in Australian radio astronomy: the Dover Heights 'hole-in-the-ground' antenna. *Journal of Astronomical History and Heritage*, 5, 21–34.

Orchiston, W., and Slee, B., 2002b. The flowering of Fleurs: an interesting interlude in Australian radio astronomy. *ATNF News*, 47, 12–15.

Orchiston, W., and Slee, B., 2005. Shame about Shain! Early Australian radio astronomy at Hornsby Valley. *ATNF News*, 55, 14–16.

Orchiston, W., and Slee, B., 2017. The early development of Australian radio astronomy: the role of the CSIRO Division of Radiophysics field stations. In Nakamura, T., and Orchiston, W., 2017. Pp. 497–578.

Orchiston, W., and Wendt, H., 2017. The contribution of the Georges Heights experimental radar antenna to Australian radio astronomy. *Journal of Astronomical History and Heritage*, 20, 313–340.

Orchiston, W., George, M., Slee, B., and Wielebinski, R., 2015a. The history of early low frequency radio astronomy in Australia. 1: The CSIRO Division of Radiophysics. *Journal of Astronomical History and Heritage*, 18, 3–13.

Orchiston, W., Lequeux, J., Steinberg, J.-L., and Delannoy, J., 2007. Highlighting the history of French radio astronomy. 3: The Würzburg antennas at Marcoussis, Meudon and Nançay. *Journal of Astronomical History and Heritage*, 10, 221–245.

Orchiston, W., Nakamura, T., and Strom, R. (eds), 2011. *Highlighting the History of Astronomy in the Asia–Pacific Region.* New York, Springer.

Orchiston, W., Slee, B., and Burman, R., 2006. The genesis of solar radio astronomy in Australia. *Journal of Astronomical History and Heritage*, 9, 35–56.

Orchiston, W., Slee, B., George, M., and Wielebinski, R., 2015b. The history of early low frequency radio astronomy in Australia. 4: Kerr, Shain, Higgins and the Hornsby Valley field station near Sydney. *Journal of Astronomical History and Heritage*, 18, 285–311.

Pawsey, J.L., 1954. Radio Astronomy Group – Current Programs. Sydney, RP. Sullivan collection.

Pawsey, J.L., 1956. Radio Astronomy Group Program. Sydney, RP. Sullivan collection.

Pawsey, J.L., 1960a. Letter to G.Z. Dimitroff, dated 19 May. RP Archives file A1/3/1.

Pawsey, J.L., 1960b. Obituary Notices. Charles Alexander Shain. *Quarterly Journal of the Royal Astronomical Society*, 1, 244–245.

Payne-Scott, R., 1947a. Possibility of constructing a spectrohelioscope or spectroheliograph suitable for investigating optical correlation with solar radio observations. Radiophysics Laboratory, Report.

Payne-Scott, R., 1947b. Solar and cosmic radio frequency radiation. Survey of knowledge available and measurements taken at Radiophysics Lab. To Dec., 1st 1945. Radiophysics, dated 21 March 1947 (Radiophysics Library: SRP 501/27).

Payne-Scott, R., 1978. Interview with Woody Sullivan, dated 3 March. Interview notes in Sullivan collection.

Radhakrishnan, V., 2006. Olof Rydbeck and early Swedish radio astronomy: a personal perspective. *Journal of Astronomical History and Heritage*, 9, 139–144.

Research Activities of the Radiophysics Laboratory, 1960. Sydney, CSIRO Division of Radiophysics.

Robertson, P., 2017. *Radio Astronomer – John Bolton and a New Window on the Universe*. Sydney, NewSouth Publishing.

Robertson, P., Orchiston, W., and Slee, B., 2014. John Bolton and the discovery of discrete radio sources. *Journal of Astronomical History and Heritage*, 17, 283–306.

Shain, C.A., 1952. Memorandum to Joe Pawsey, dated 6 February. RP Archives file A1/3/1(a).

Shain, C.A., 1958. The Sydney 19.7 Mc/s radio telescope. *Proceedings of the Institute of Radio Engineers*, 46, 85–88.

Sheridan, K.V., 1978. Taped interview with Woody Sullivan, 7 March.

Shouguan, W., 2017. The early development of Chinese radio astronomy: the role of W.N. Christiansen. In Nakamura, T., and Orchiston, W., 2017. Pp. 245–254.

Sim, H., 2013. Vale R.X. McGee (1920–2012). *ATNF News*, 74, 17–18.

Slee, O.B., 1978. Taped interview with Woody Sullivan, 1 March.

Slee, O.B., 1994. Some memories of the Dover Heights field station, 1946–1954. *Australian Journal of Physics*, 47, 517–534.

Slee, O.B., 2005. Early Australian measurements of angular structure in discrete radio sources. *Journal of Astronomical History and Heritage*, 8, 97–116.

Stanley, G.J, 1986. Letter to Woody Sullivan, dated 23 April. In Sullivan collection.

Stanley, G.J., 1994. Recollections of John Bolton at Dover Heights and Caltech. *Australian Journal of Physics*, 47, 507–516.

Stanley, G.J., 1997. Letter to Miller Goss, dated 9 October. Copy in Sullivan collection.

Stewart, R.T., 2009. The Contribution of the CSIRO Division of Radiophysics Penrith and Dapto Field Stations to International Radio Astronomy. PhD thesis, James Cook University, Queensland.

Stewart, R., Orchiston, W., and Slee, B., 2011a. The contribution of the Division of Radiophysics Dapto field station to solar radio astronomy, 1952–1964. In Orchiston, W., et al., 2011. Pp. 481–526.

Stewart, R., Orchiston, W., and Slee, B., 2011b. The Sun has set on a brilliant mind: Paul Wild (1923–2008). In Orchiston, W., et al., 2011. Pp. 527–542.

Stewart, R., Wendt, H., Orchiston, W., and Slee, B., 2010. The Radiophysics field station at Penrith, New South Wales, and the world's first solar radio spectrograph. *Journal of Astronomical History and Heritage*, 13, 2–15.

Sullivan, W.T. (ed.), 1984. *The Early Years of Radio Astronomy. Reflections Fifty Years after Jansky's Discovery*. Cambridge, Cambridge University Press.

Sullivan, W.T., 2009. *Cosmic Noise: A History of Early Radio Astronomy*. Cambridge, Cambridge University Press.

Swarup, G., 2006. From Potts Hill (Australia) to Pune (India): The journey of a radio astronomer. *Journal of Astronomical History and Heritage*, 9, 21–34.

Swarup, G., 2008. Reminiscences regarding Professor W.N. Christiansen. *Journal of Astronomical History and Heritage*, 11, 194–202.

Thompson, A.R., 2010. The Harvard radio astronomy station at Fort Davis, Texas. *Journal of Astronomical History and Heritage*, 13, 17–27.

Treharne, R.F., 1983. Multipurpose whole-band HF antenna architecture. *Journal of Electrical and Electronics Engineering, Australia*, 3, 141–152.

Van Woerden, H., and Strom, R., 2006. The beginnings of radio astronomy in the Netherlands. *Journal of Astronomical History and Heritage*, 9, 3–20.

Wendt, H., 2008. The Contribution of the Division of Radiophysics Potts Hill and Murraybank Field Stations to International Radio Astronomy. PhD thesis, James Cook University, Queensland.

Wendt, H., and Orchiston, W., 2018. The contribution of the AN/TPS-3 radar antenna to Australian radio astronomy. *Journal of Astronomical History and Heritage*, 21, 65–80.

Wendt, H., and Orchiston, W., 2019. The short-lived CSIRO Division of Radiophysics field station at Bankstown Aerodrome in Sydney. *Journal of Astronomical History and Heritage*, 22, 266–272.

Wendt, H., Orchiston, W., and Slee, B., 2008a. The Australian solar eclipse expeditions of 1947 and 1949. *Journal of Astronomical History and Heritage*, 11, 71–78.

Wendt, H., Orchiston, W., and Slee, B., 2008b. W.N. Christiansen and the development of the solar grating array. *Journal of Astronomical History and Heritage*, 11, 173–184.

Wendt, H.W., Orchiston, W., and Slee, B., 2008c. W.N. Christiansen and the initial Australian investigation of the 21-cm hydrogen line. *Journal of Astronomical History and Heritage*, 11, 185–193.

Wendt, H., Orchiston, W., and Slee, B., 2011a. The contribution of the Division of Radiophysics Potts Hill field station to international radio astronomy. In Orchiston, W., et al. 2011. Pp. 379–431.

Wendt, H., Orchiston, W., and Slee, B., 2011b. The contribution of the Division of Radiophysics Murraybank field station to international radio astronomy. In Orchiston, W., et al., 2011. Pp. 433–479.

Wendt, H., Orchiston, W., and Slee, B., 2011c. An overview of W.N. Christiansen's contribution to Australian radio astronomy, 1948–1960. In Orchiston, W., et al., 2011. Pp. 547–587.

Westfold, K.C., 1994. John Bolton – some early memories. *Australian Journal of Physics*, 47, 535–539.

Wild, J.P., 1951. Mcmorandum to J.L. Pawsey dated 1 January. Sullivan collection.

Wild, J.P., 1978. Taped interview with Woody Sullivan, 3 March.

Wild, J.P., and McCready, L.L., 1950. Observations of the spectrum of high-intensity solar radiation at metre wavelengths. Part 1. The apparatus and spectral types of solar bursts observed. *Australian Journal of Scientific Research*, A3, 387–398.

Wild, J.P., and Radhakrishnan, V., 1995. John Gatenby Bolton, 5 June 1922 – 6 July 1993. *Biographical Memoirs of Fellows of the Royal Society*, 41, 72–86.

Wild, J.P., and Sheridan, K.V., 1958. A swept-frequency interferometer for the study of high-intensity solar radiation at metre wavelengths. *Proceedings of the Institute of Radio Engineers*, 46, 160–171.

Chapter 3
Exploring the Neighbourhood – the Sun, the Moon and Jupiter

3.1 The Sun

The idea that the Sun may emit radio waves has a long history. As Woody Sullivan (1982) has shown in his *Classics in Radio Astronomy*, the idea first gained credence in the 1890s, but initial searches by a number of different investigators proved fruitless (e.g. see Débarbat et al. (2007) for details of Nordmann's search in 1901). Success would come half-century later, thanks largely to the development of wartime radar.

Solar radio astronomy had its foundations during WWII when independent detections of solar emission were made in the USA, Denmark (by German forces), England, Australia and New Zealand. Because of security issues this work was largely classified until after the war, but Joe Pawsey (1908–1962) must have been aware of some of these findings because in 1944 he carried out a very brief abortive attempt to detect solar radio emission from the Radiophysics Lab in Sydney. According to Mills (1920–2011; 1976), "… he stuck a simple parabolic reflector out of the window and looked at the Sun and saw nothing (because it was at 10 cm)." This was a hurried, solitary, opportunistic experiment, and it was not followed up.

Then in mid-1945 almost simultaneously the Radiophysics (henceforth RP) group received secret accounts of the independent discovery of solar radio emission at New Zealand and British radar stations, plus a copy of the 1944 *Astrophysical Journal* paper by American Grote Reber (1911–2002) reporting his detection of solar radio waves (see Fig. 3.1). It was these discoveries that collectively precipitated the Australian 'solar noise' program. The Sydney group (Fig. 3.2) would quickly gain world supremacy in solar radio astronomy, and maintain this position through until the mid-1980s when the closing of the Culgoora Radioheliograph and staff promotions and retirements saw the demise of the RP Solar Group. See Orchiston et al. (2006) for the genesis of solar radio astronomy in Australia; see also Stewart et al. (2011b) for an overview of Australian solar radio astronomy 1945 to 1960, and Stewart et al. (2011a) for Paul Wild's impressive contribution to Australian

© The Author(s) 2021
W. Orchiston et al., *Golden Years of Australian Radio Astronomy*,
Historical & Cultural Astronomy, https://doi.org/10.1007/978-3-319-91843-3_3

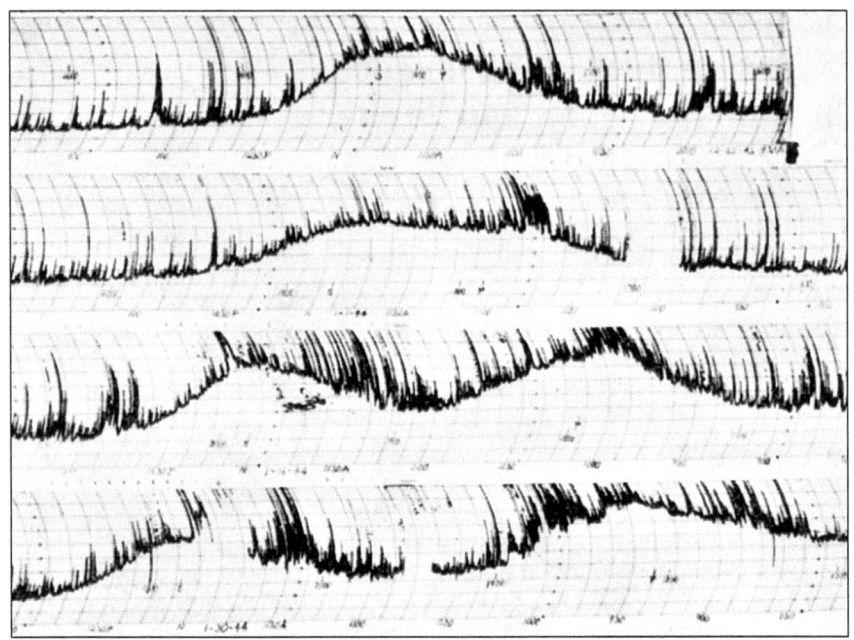

Fig. 3.1 Grote Reber's observations at 160 MHz, showing solar bursts superimposed on Galactic radio emission (after Reber, 1944: 286)

and international solar radio astronomy. Sullivan (2009) covers not only Australian solar research in the post-war decade, but places it in the context of other research done around the world.

3.1.1   The first solar observations at Radiophysics

The first serious Sydney observations were inspired by Dr Elizabeth Alexander (1908–1958; see Fig. 3.3 and Biobox 3.1) who was responsible for analysing and explaining observations of solar bursts made at 200 MHz in neighbouring New Zealand and on Norfolk Island (in the Tasman Sea between Australia and New Zealand). The future head of the RP radio astronomy group, Joe Pawsey (Fig. 1.13), noticed that Australia had similar 200 MHz COL (chain overseas low-flying) radar antennas deployed at RAAF radar stations around the country, so he arranged access to the antenna at Collaroy, a coastal suburb of Sydney, in an attempt to replicate Alexander's observations.

From 3 October 1945 RAAF personnel at Station 54 (see Fig. 2.1) agreed to carry out daily solar monitoring on behalf of RP for a 75 minute interval beginning about 10 minutes before sunrise, and this brought instant results:

Fig. 3.2 A Culgoora Radioheliograph antenna and members of the Radiophysics Solar Group during the mid-1960s. Key, inner circle, clockwise from top, centre Paul Wild (Group Leader), Masaki Morimoto, Charlie Attwood, John Sparks, Jack Palmer, Bill Bowie, Les Clague, Geoff Chandler and Kevin Sheridan; outer circle, Steve Smerd, Shigamasa Suzuki, Ron Stewart, Gopala Rao, Len Binskin, Joe Mack, Wayne Orchiston, Nick Fourikis, Warren Payten, Norman Labrum and Alan Weiss. Absent are T. Krishnan, Keith McAlister, Don McLean and Dick Mullaly [courtesy: CSIRO Radio Astronomy Image Archive (CRAIA)]

> We observed, from the direction of the Sun, a considerable amount of radiation having the apparent characteristics of fluctuation 'noise' when observed on a cathode-ray oscillograph or head-phones. However, the output meter reading fluctuated considerably, a characteristic which is not typical of normal thermal agitation "noise". (Pawsey et al., 1946).

Alexander also had concluded that the emission observed previously from New Zealand was non-thermal in origin. Joe Pawsey, Ruby Payne-Scott (1912–1981) and Lindsay McCready (1910–1976) were especially interested in any possible relationship between solar radio emission and optical features, but particularly sunspots, and when these parameters were plotted for the period 3–23 October 1945 a general correlation was apparent. This confirmed the claims made earlier by Stanley Hey (1909–2000) and Alexander. Pawsey, Payne-Scott and McCready (1946) reported their results in the 9 February 1946 issue of *Nature*, shortly after Hey's belated announcement of his wartime observations (Hey, 1946).

Fig. 3.3 Dr Elizabeth Alexander (1908–1958), the world's first female radio astronomer, moved in the upper echelons of New Zealand science (from left): Dr (later Sir) Ernest Marsden (Director of Scientific Developments, DSIR), Sir Cyril Newall (Governor-General of New Zealand), Alexander, Dr Ian Stevenson (Director of the Radio Development Laboratory) and the Governor-General's Aide de Camp (courtesy: Professor E.R. Collins)

Biobox 3.1: Elizabeth Alexander

Francis Elizabeth Somerville Alexander (neé Caldwell) (Fig. 3.4) was born in Merton, England on 13 December 1908 and the following year moved to India where her father was a college professor. She returned to England for her secondary schooling, and then studied geology and physics at Cambridge graduating with a PhD in geology in 1934. Following her marriage Elizabeth moved to Singapore in 1936 where she worked with the Royal Navy on radio direction finding (later known as radar). In 1942, just before the Japanese occupation of Singapore, she and her three young children escaped to New

Fig. 3.4 Dr Elizabeth Alexander (1908–1958) was the world's first female radio astronomer (courtesy: Mary Harris)

(continued)

Biobox 3.1 (continued)
Zealand, where she was appointed head of the Operational Research Section of the Radio Development Laboratory.

At the Laboratory Elizabeth was involved in radar development, but also found time to investigate solar radio emission – the mysterious 'Norfolk Island Effect'. At the end of the war, after her husband was released from a Japanese prisoner-of-war camp, they returned to Singapore in 1947. Elizabeth served for a period as Registrar of the University of Malaya and then became the Government Geologist for Singapore. In 1952 her family moved to Nigeria where Elizabeth died of a stroke in October 1958, shortly before her 50th birthday. Elizabeth Alexander was the world's first woman to work in the field that would later be known as radio astronomy. After leaving New Zealand she never conducted any further astronomical studies, although she did publish a short paper on the 'Norfolk Island Effect' in 1946. Further details of Elizabeth's life are provided in a captivating book written by her daughter, Mary Harris (2017), and in papers by Orchiston (2005, 2016).

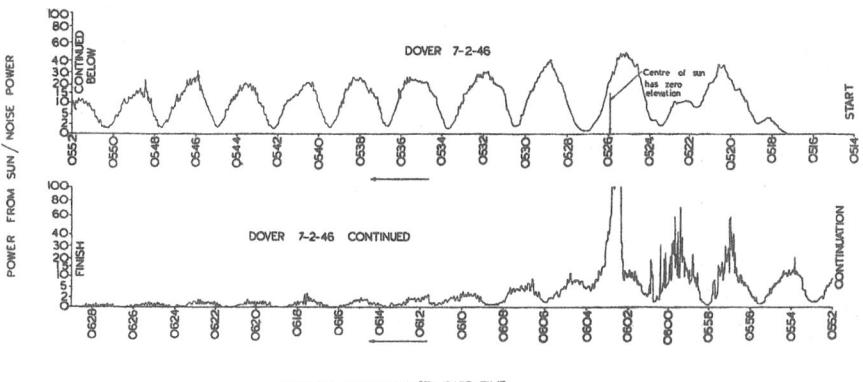

Fig. 3.5 Chart record of the Sun passing through the sea interferometer fringes at Dover Heights on 7 February 1946. Time, in minutes, runs from right to left, and the chronological sequence is continued in the lower strip where intense burst emission is apparent just before the Sun rises above the beam of the aerial (after McCready et al., 1947: 366)

Solar monitoring at Collaroy Plateau was continued until March 1946, and soon was joined by parallel observations made with the antenna at the Dover Heights radar station (Fig. 2.1), about 17 km south of Collaroy. In order to obtain a visual record of solar radio emission, chart recorders were included in the receiving systems, and these provided a permanent record of the Sun's passage through the sea interferometer fringes, not only revealing temporal variations in the background level of solar radiation but also intense bursts of short duration (see Fig. 3.5).

Meanwhile, McCready, Pawsey and Payne-Scott (1947: 363) continued to investigate the general relationship between sunspots and solar emission. By plotting the general background level of solar radiation from October 1945 to February 1946 against sunspot number and sunspot areas, the correlation noted earlier for October 1945 was confirmed, but in this instance they found that "… the correlation with areas is somewhat closer than with sunspot number but neither is exact." The Sun was particularly active during this period, and as Sullivan (1982: 183) has pointed out, "… these novice astronomers were somewhat lucky in being able to observe the great sunspot group of February, 1946, one whose main sunspot is amongst the largest ever photographed."

McCready and his colleagues also investigated the location of the source of emission on the Sun by analysing the interference fringes in Fig. 3.5 and, during the presence of the great sunspot group of February 1946, were able to demonstrate conclusively that the solar radiation originated from a strip that in each case included this sunspot group. Examples deriving from Collaroy and Dover Heights are shown in Fig. 3.6, and

> In each case the radiating strip has a width considerably less than that of the Sun's disk, being of the order of the size of the sunspot group, and passes through the group. It moves across the Sun with the spots as the Sun rotates … There seems no reasonable doubt that the source was localised in a small region in the vicinity of the spots. [However] The observations do not provide any information as to the detailed structure of the source within this region. (McCready et al., 1947: 368).

Although most of these pioneering observations were made at 200 MHz, a few observations were also carried out at 75 and 3000 MHz. At the higher frequency the low level of solar radiation detected was consistent with results George Southworth (1890–1972) published in 1945, whereas at 75 MHz the solar emission was comparable to that recorded at 200 MHz.

In addition to burst emission, Pawsey was interested in the quiescent level of solar radiation at 200 MHz, and by isolating and quantifying the non-burst component was able to demonstrate the existence of a 'hot corona', something that had been hinted at previously by optical astronomers. The announcement by Pawsey (1946) in *Nature* of a temperature of one million degrees immediately followed a theoretical paper by David Martyn (1906–1970) predicting a coronal temperature of this order. Originally the inaugural chief of the Radiophysics Lab, Martyn was at that time based at the Commonwealth Solar Observatory, near Canberra (Biobox 3.2).

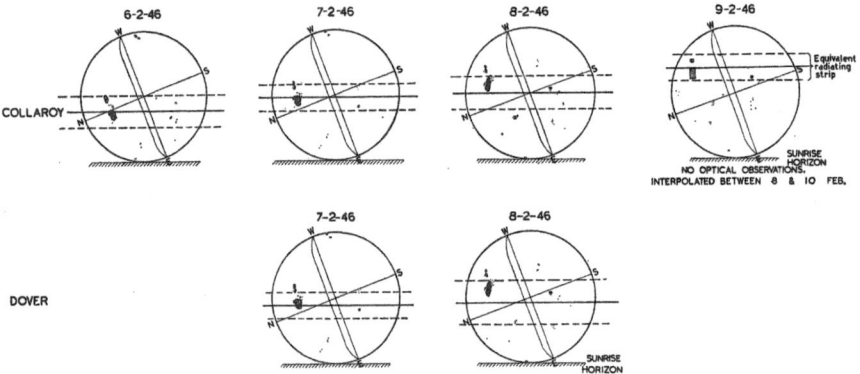

Fig. 3.6 Drawings of the Sun made between 6 and 9 February 1946 showing the position and width of the 'equivalent radiating strips' from which the solar radiation originated (after McCready et al., 1947: 369)

Biobox 3.2: David Martyn
David Forbes Martyn (Fig. 3.7) was born in Cambuslang, near Glasgow, Scotland, in 1906, and died in Camden (near Sydney) in 1970. After completing BSc and PhD degrees at Imperial College and the University of London in 1926 and 1928, respectively, he was appointed a Research Officer with the Radio Research Board in Australia and moved to Melbourne. He then published a series of research papers on the ionosphere, most notably one in 1936, co-authored by Owen Pulley (1906–1966). Of this paper Piddington and Oliphant (1972: 49) later wrote: "It is no exaggeration to say that this paper, of itself, trans-formed all subsequent thinking on almost all aspects of the high atmosphere."

In 1939 Martyn was selected to head the CSIR's new Division of Radiophysics, then involved in the wartime development of radar. This did not prove a successful appointment and in March 1941 he was replaced by Fred White (1905–1994), while David was appointed the head of a new Operational Research Group. By the end of the War, Martyn was based at Mount Stromlo Observatory, where he continued his ionospheric research and wrote a number of seminal papers on solar radio emission. Although still on the CSIRO payroll, David remained at Mount Stromlo until 1958, when he took charge of the CSIRO's new Upper Atmosphere Section at Camden, near Sydney.

A Fellow of the Royal Society since 1950, Martyn received many awards. He served as President of the Radio Astronomy Commission and later the Ionospheric Commission of the International Union of Radio Science. David also chaired a United Nations sub-committee on the Peaceful Uses of Outer Space and was on the Executive Committee of the International Council of Scientific Unions. Martyn was largely responsible for the forma-tion of the Australian Academy of Science in the early 1950s – he was its inaugural Secretary (Physical Sciences) and was its President at the time of his death (Orchiston, 2014a).

Fig. 3.7 David Martyn was the inaugural Chief of the Radiophysics Lab (courtesy: Mount Stromlo and Siding Spring Observatories)

While these pioneering observations were under way in Sydney, a young radar mechanic Bruce Slee (1924–2016) was busy carrying out his own solitary solar observations at an RAAF radar station near Darwin, in total ignorance of this new RP research initiative. Station 59 at Lee Point featured a British 200 MHz COL Mk5 radar unit situated about 500 m from the coast and mounted atop a 46 m high steel tower. Between October 1945 and March 1946 Slee carried out observations that were an independent discovery of solar radio emission. From time to time he noticed that in the hour leading up to sunset

> ... the "grass" on the range display increased its height by up to a factor of ten when the antenna was pointing towards the setting Sun. By slowly scanning backwards and forwards through the Sun, he was able to establish that the source of the signal lay at the solar azimuth to within the errors of measurement. Furthermore, when he stopped the antenna while pointing at the Sun, he noticed that the amplitude of the grass varied regularly by a large factor with a period of about 3 minutes. He concluded that this behaviour was consistent with the setting Sun passing through the sea interference fringes formed by the antenna and its image in the sea. (Orchiston and Slee, 2002: 27).

In March 1946, Slee read a newspaper account of the RP solar research and wrote a letter to Pawsey describing his own work. This drew an enthusiastic response, and a collaborative program of solar monitoring was arranged, only to be immediately stymied by the closing of the Lee Point radar station. In 1947 a tender of £1,200 was accepted for removal of the tower, and another phase in the wartime history of Australian radar came to an end. But at the same time Slee was enjoying a new career at RP in far-off Sydney, and he would go on to acquire a Doctor of Science degree and to build an international reputation as a radio astronomer (for biographical details see Orchiston, 2004a and Biobox 4.1).

One of the major problems with radar antennas – including those at Collaroy and Dover Heights – was that they could not track the Sun, so in early 1946 RP staff installed simple, movable 200 MHz Yagi-type antenna arrays (Fig. 2.13) at Dover Heights, at the North Head radar station (located at the entrance to Sydney Harbour – see Fig. 1.14), and at Mount Stromlo Observatory (see Fig. 2.8). Associated with the 4-Yagi array at Mount Stromlo was Cla Allen (1904–1987; Biobox 3.3), who specialised in solar spectroscopy and the terrestrial effects of solar flares. However, his wartime research into the causes of short-wave radio fadeouts whetted his appetite to investigate solar radio emission, and this fitted well with the desire of the Mount Stromlo Director, Richard Woolley (1906–1986; Fig. 1.36), to include an active radio astronomy program in the Observatory's post-war research portfolio.

Between April 1946 and March 1947 Allen (1947) carried out solar monitoring and found that even on radio-quiet days

> ... there has always been a detectable amount of radiation which appears to be quite variable ... [In addition] there are occasional sudden "bursts" of solar radio-noise which last for periods of the order of 1 sec ... [and] rather rarely, sudden outbursts of radio noise, which last for a few minutes, fluctuating violently, and then disappear.

Allen confirmed that solar emission was closely related to the central meridian passage of sunspots, just as Pawsey et al. (1946) had reported, although not all sunspots produced solar noise. Meanwhile, his analysis failed to show a general correlation

Biobox 3.3: Cla Allen

Clabon ('Cla') Walter Allen (Fig. 3.8) was born in Perth, Western Australia, in 1904, and obtained BSc, MSc and DSc degrees from the University of Western Australia in 1926, 1929 and 1936 respectively. In 1926 he was appointed as one of the founding astronomers of the Commonwealth Solar Observatory near Canberra (later named Mount Stromlo Observatory). During the 1930s Cla mapped the solar spectrum at optical wavelengths and carried out a variety of atmospheric research projects.

Following the launch of the RP solar radio astronomy program in late 1945, Allen arranged for a 200 MHz four-element Yagi array (see e.g. Fig. 2.13) to be installed at Mount Stromlo. He used this radio telescope to monitor solar burst emission, and for an all-sky survey of Galactic radio emission. In 1951 Allen accepted the newly-created Perrin Chair of Astronomy at University College London, where he wrote *Astrophysical Quantities*, a book for which he is justly famous. Irish astronomer Derek McNally (1934–2020; 1990: 265) described Allen as "… a man of great personal integrity, sincerity, determination and modesty. He did not seek personal aggrandisement and devoted his energies to his research, his students and his department." After retiring in 1972 Cla returned to Canberra, where he had an honorary position at Mount Stromlo. He died in 1987.

Fig. 3.8 Cla Allen at Mount Stromlo Observatory was a strong supporter of the solar research at Radiophysics (courtesy: Mount Stromlo and Siding Spring Observatories)

between solar noise and hydrogen alpha line (Hα) features or geomagnetic storms, although some solar flares were associated with outbursts.

While Allen was conducting these investigations, his Mt Stromlo colleague David Martyn was using data obtained with the 200 MHz array to research the polarisation of solar radio emission. Martyn (1946) reasoned that since solar bursts were associated in some way with sunspots and in turn sunspots were associated with strong magnetic fields, "… we should expect to find evidence of the magnetic field in the production of gyratory effects at the source of the [radio] emissions, and/ or in differential absorption of right-handed and left-handed components of polarisation during transmission through the corona."

The passage of the large sunspot group of July 1946 provided an ideal opportunity for Martyn to test this hypothesis, by turning two of the Yagis in the array at right-angles to the other two Yagis. Observations made on July 26 revealed that "… the right-handed circularly polarised power received was some seven times greater than that received when the system accepted only left-handed circularly polarised radiation." (Martyn, 1946). This result was mirrored during observations at 60 and 100 MHz made by John Bolton (1922–1993) and Gordon Stanley (1921–2001) at Dover Heights in March and April 1947.

Martyn followed up this important paper with a theoretical contribution in which he proposed the revolutionary idea that the solar corona, the tenuous outer atmosphere of the Sun, has a temperature of about one million degrees, far higher than the 6000 degrees that characterises the photosphere where sunspots are found. After contributing two further theoretical papers, Martyn moved on to other research interests, and when Allen accepted a professorship at University College London, in 1951, Mount Stromlo's involvement in radio astronomy came to an abrupt end.

At about the same time that Allen was beginning his solar radio astronomy research at Mount Stromlo, Sydney Williams (1910–1979), a lecturer in physics at the University of Western Australia, set up a 75 MHz equatorially-mounted Yagi antenna on campus in suburban Perth. Williams had a background in optical astronomy at the Commonwealth Solar Observatory, and his three-year foray into radio astronomy was apparently inspired by what he heard at a seminar held at the Radiophysics Lab in January 1946. He was particularly interested in temporal variations in the intensity of solar radio emission, correlations with sunspots, solar flares and ionospheric radio fadeouts, and the shapes of short pulses of solar radio emission.

Parallel daily monitoring with these steerable antennas in Perth, at Mount Stromlo and in Sydney revealed almost identical patterns of solar emission, with a changing level of background radiation upon which were superimposed bursts of varying duration and intensity. The precise correspondence in the case of bursts, and the fact that burst activity did not vary systematically as the Sun rose from the horizon towards the zenith, proved conclusively that the emission was of solar origin and not caused by ionospheric scintillations.

The next challenge was to expand these multi-frequency observations, with particular emphasis on burst emission, and from mid-1946 the relative arrival times of bursts were recorded at 60, 75 and 200 MHz at Dover Heights. In 1947, Payne-Scott, Don Yabsley (1923–2003) and Bolton reported on their research in a short paper published in *Nature*. They found that small bursts often were not correlated at the three frequencies, whereas many of the larger bursts were present at all three frequencies, but did not occur simultaneously. Instead, they arrived in the sequence 200, 75 and 60 MHz, with a typical delay of about 2 s between bursts at 200 and 75 MHz, and a similar interval between bursts at 75 and 60 MHz. A few observations were also made at 30 MHz, and the delay in arrival times of bursts at 60 and 30 MHz was also a few seconds.

Very much rarer were major 'outbursts', which could last for hours, and in this instance the delays in the respective arrival times at the different frequencies were of the order of several minutes rather than seconds. An outburst recorded on 7 March 1947 was associated with a solar flare and short-wave radio fadeout, and an aurora was visible from some areas of Australia on the evening of 9 March. In interpreting their observations, Payne-Scott, Yabsley and Bolton (1947) concluded that "The successive delays between the onset of the outburst on 200, 100 and 60 Mc/s. suggest that the outburst was related to some physical agency passing from high-frequency to lower-frequency levels [in the solar corona]." This spectacular event would later be classed as a Type II burst, and its discovery played a crucial role in the subsequent development of solar radio astronomy.

3.1.2   Solar eclipses and solar bursts

At the end of 1946 Gordon Stanley began building radio equipment to observe a total solar eclipse in Brazil. The previous year, the Canadian radio astronomer Arthur Covington (1913–2001) had demonstrated that eclipses could be used to determine the positions of centres of radio emission in the solar corona, and this offered an excellent opportunity to look at their association with optically-active regions. In the end this project was torpedoed by "... the appalling difficulties in getting transport of men and equipment to Brazil." (Bowen, 1947), and the equipment was transferred to Dover Heights where it was used in conjunction with other antennas by Stanley and Bolton.

By August 1947, Georges Heights had joined Dover Heights as a solar radio astronomy field station. One of the three ex-radar antennas there (see Fig. 2.25) was brought into operation, and solar monitoring was carried out for about two hours daily, from 18 August until 30 November. This resulted in the detection of many bursts at 200 MHz. In contrast, bursts were rare at 600 and 1200 MHz, where the general flux variations with time were correlated with sunspots (Fig. 3.9). This distinctive pattern was discussed by RP's Fred Lehany (1915–1980) and Don Yabsley in papers published in *Nature* and the *Australian Journal of Scientific Research* (Lehany and Yabsley, 1948, 1949), along with those rare outbursts that would later be classed as Type II events.

During the first nine months of 1948, Payne-Scott studied solar emission at 18.3, 19.8, 60, 65 and 85 MHz at the Hornsby Valley field station, and detected many correlated bursts. In addition to the three different kinds of bursts noted by earlier workers, she also identified a fourth type of event: "The intensity reaches a high level and remains there for hours or days on end; there are continual fluctuations in intensity, both long-term and short-term ... This type of radiation will be called "enhanced radiation" ... [and] Superimposed on it may be bursts." (Payne-Scott, 1949: 216–17). An example of enhanced radiation at 60 and 85 MHz is shown in Fig. 3.10. Payne-Scott found this enhanced emission was circularly polarised, and was usually associated with the presence of large sunspot groups.

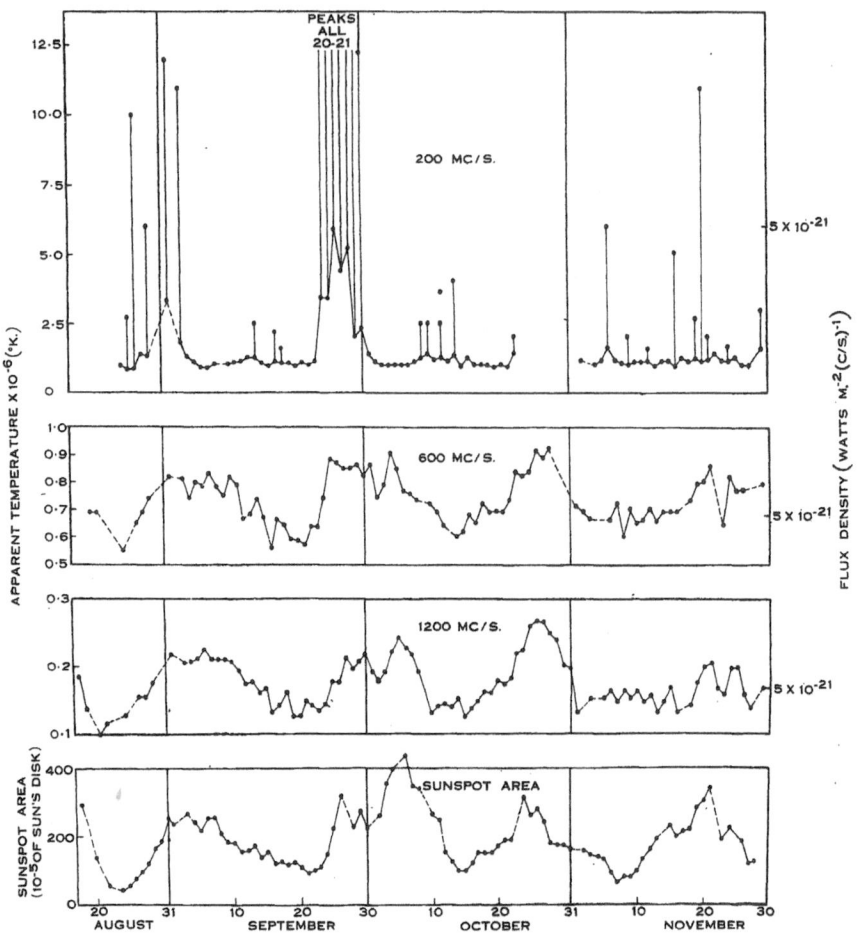

Fig. 3.9 The correlation between sunspot area and solar flux at 200, 600 and 1200 MHz (after Lehany and Yabsley, 1949)

So by the end of 1948 it was clear that the Sun produced 'enhanced radiation' – which typically occurred with small-scale bursts of short-duration and restricted in frequency – and at least two different types of major burst; larger relatively short-lived bursts which were represented over a range of frequencies and had starting times that varied by just a few seconds as one moved down in frequency; and long-lasting major outbursts, which were also present over a range of frequencies, but exhibited starting-time delays of several minutes from one frequency to the next. A

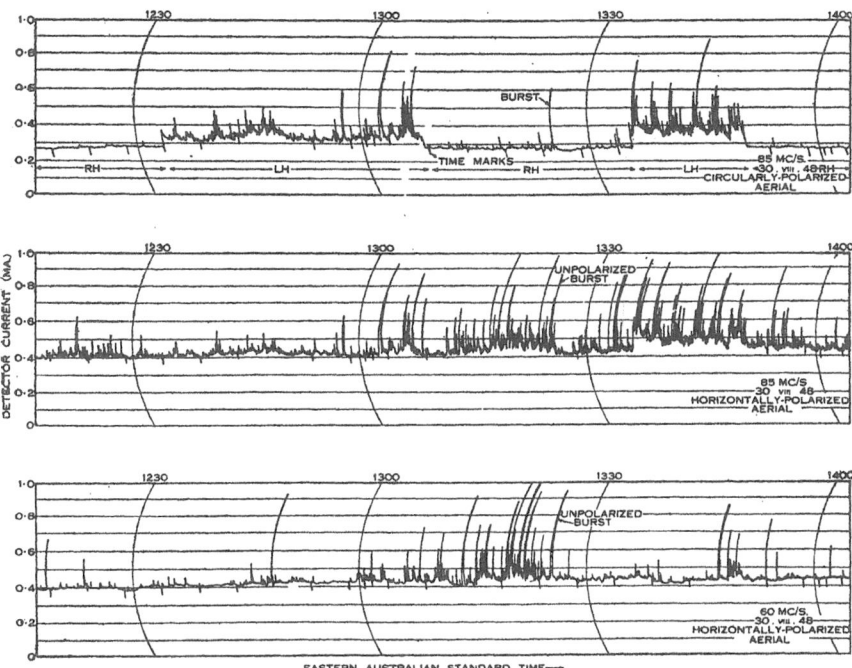

Fig. 3.10 Solar bursts and 'enhanced emission' recorded at 60 and 85 MHz (after Payne-Scott, 1949: 218)

magnificent example of this last type of solar burst is Fig. 3.11. This led naturally to two quite different types of research: investigation of the spectral features of the burst emission, and examination of the location, movement and polarisation of the bursts.

Determining the positions of the solar regions generating radio emission was a major challenge given the poor angular resolution of the radio telescopes available in the late 1940s. McCready, Pawsey and Payne-Scott had already shown how sea interferometer observations could reveal the radiating strip associated with burst emission (Fig. 3.6), but this did not give an unambiguous two-dimensional position. However, a solar eclipse could provide this, if observed simultaneously from two or more widely-spaced sites. In the course of the eclipse, as the Moon progressed across the solar disk it masked different regions of the Sun's atmosphere at different times. As seen from the different observing sites, the Moon's limb followed different paths, and by noting the times when different optically-active regions were masked and exposed again the sources of any associated radio emission could be pinpointed.

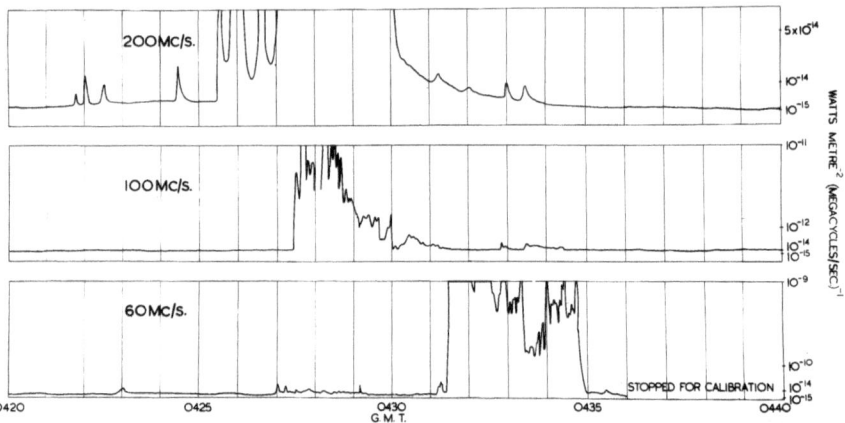

Fig. 3.11 The large outburst recorded at three different frequencies on 8 March (after Payne-Scott et al., 1947: 256). In 2020, the 1947 outburst at 60 MHz is still the strongest ever recorded

The partial solar eclipse of 1 November 1948 provided an ideal opportunity to exploit this method, and so observations were carried out at 600, 3000 and 9400 MHz from the Potts Hill field station and at 600 MHz from two remote sites, Rockbank, just north–west of suburban Melbourne (Fig. 2.8), and Strahan, on the west coast of Tasmania (Orchiston, 2004b). Collectively, these observations involved four different teams of scientists: W.N. ('Chris') Christiansen (1913–2007) and Bernie Mills (1920–2011) (Christiansen et al., 1949); Don Yabsley and John Murray; Harry Minnett (1917–2003) and Norman Labrum (1921–2011); and Jack Piddington (1910–1997) and Jim Hindman (1919–1999). Interestingly, all except Christiansen and Labrum would go on to build reputations in *non*-solar areas of radio astronomy.

The Rockbank eclipse results are representative, and can be used to illustrate the type of analysis involved. At the time of the eclipse, visual monitoring in Sydney revealed the presence of six different groups of sunspots. Although only 72% of the Sun's disk would be covered at the peak of the eclipse, it was noted that as the eclipse progressed the declining level of radio emission was punctuated by a succession of small peaks and troughs (Fig. 3.12). The troughs occurred when centres responsible for enhanced radio emission were covered, and their precise positions were obtained by plotting the intersections of the Sydney, Strahan and Rockbank lunar-limb paths across the Sun during the eclipse. These locations are shown in Fig. 3.13. Of the eight centres of enhanced solar emission, three coincided with the positions of sunspot groups, three others were close to positions occupied by sunspot groups exactly one solar rotation earlier, and a seventh was located about 170,000 km off the south–west limb of the Sun, directly above a magnetically-active region associated with a limb prominence. However, two small sunspot groups and one large group were not associated with measurable levels of radio radiation, showing that there was not always an exact correlation between sunspots and solar emission.

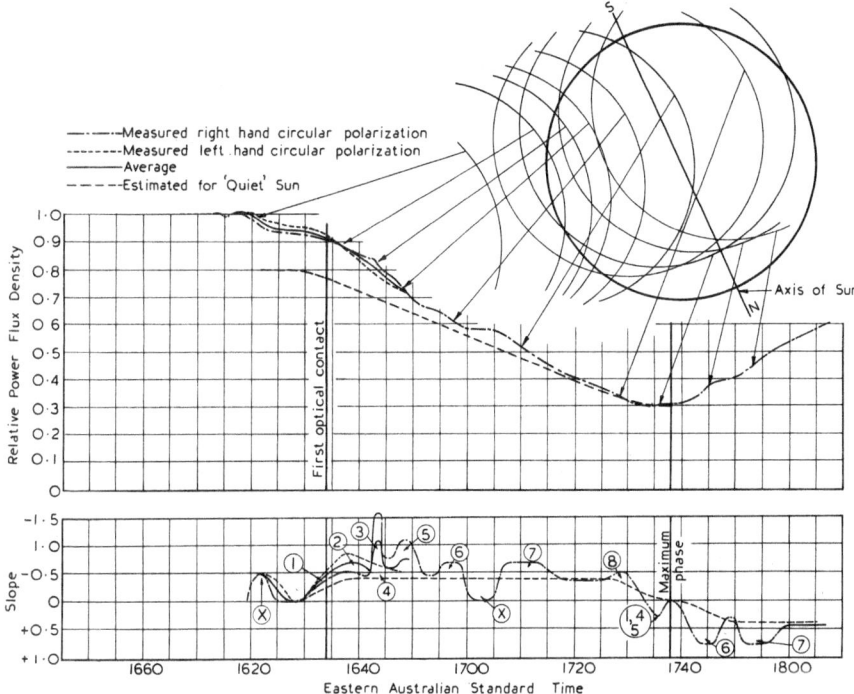

Fig. 3.12 Variations in 600 MHz radio emission recorded at Rockbank (Victoria) during the 1 November 1948 partial solar eclipse. The upper plot shows the actual chart record, while the lower plot has been corrected for the slope and the peaks have been magnified (after Christiansen et al., 1949: 511)

These eclipse observations confirmed the existence of two discrete components of non-burst solar emission: a basic component of thermal origin, which originates from the whole disk of the Sun, and a 'slowly-varying component' (first recognised by Jean-Françoise Denisse (1949) in France) that is generated in small localised regions that are often associated with sunspots. This emission was discussed in detail by Pawsey and Yabsley and by Piddington and Minnett, in papers that were published in the *Australian Journal of Scientific Research* in 1949 and 1951 respectively (see also Pawsey, 1950).

There was another partial solar eclipse on 22 October 1949 that also was visible from Australia and, primed by the success of their 1948 campaign, RP mounted an equally ambitious program with radio telescopes sited in Sydney (at Potts Hill), Victoria and Tasmania (but at different sites to those used the previous year). On this occasion the selected observing frequency was 1200 MHz, while Cla Allen carried out parallel observations at 200 MHz from Mt Stromlo. Although observations of this eclipse also were successful, for reasons suggested by Wendt et al. (2008) no publications resulted from this particular campaign. American war-surplus AN/ TPS-3 antennas were used at remote sites during these eclipse expeditions, and their

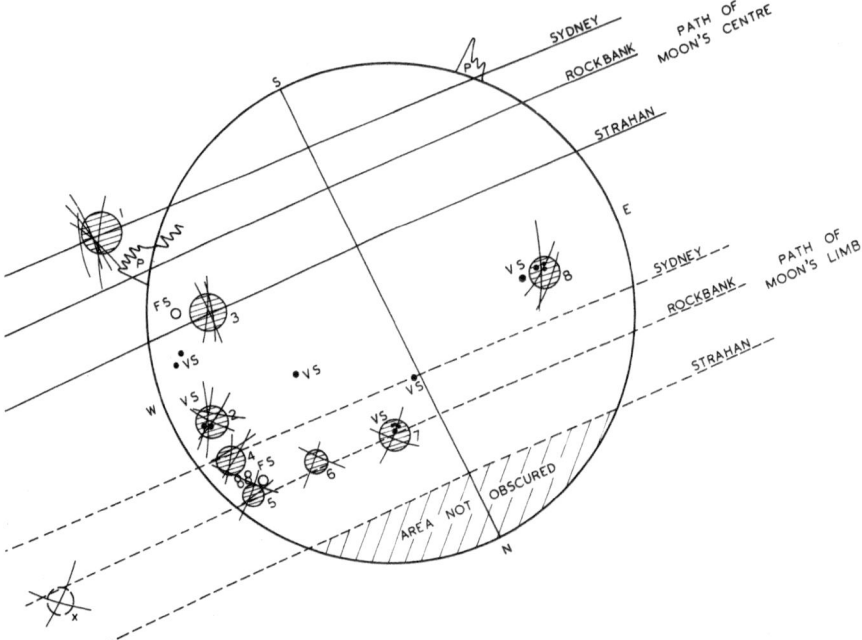

Fig. 3.13 Regions of enhanced solar emission at 600 MHz (hatching) recorded during the partial solar eclipse of 1 November 1948 shown in relation to locations of optical activity: VS = visible sunspots; FS = positions of sunspots during the previous solar rotation; and P = prominences (after Christiansen et al., 1949: 513)

significant role in early Australian radio astronomy is discussed by Wendt and Orchiston (2018).

While eclipses can tell us about the locations of solar emission, they are rare events, and as Paul Wild (1923–2008) reminded us, "… the analysis is complex and the result often inconclusive." (Wild, 1953: 687). There had to be better ways of investigating the positions of these sources, and Ruby Payne-Scott and Alec Little (1925–1985) came up with an ingenious method (Payne-Scott and Little, 1951). Using 97 MHz swept-lobe and fixed-lobe interferometers set up at the Potts Hill field station (see Fig. 2.39), they were able to measure the positions and polarisation of sources of burst emission (Fig. 3.14). Between May 1949 and August 1950 observations were made on week days for about two hours either side of meridian transit of the Sun, and at times of high solar activity also on weekends. Records were obtained of 30 noise storms, 6 outbursts and 25 randomly-polarised bursts.

They found an association between noise storms and sunspots, but only with the largest spot in any group. Furthermore, sunspots were linked to strong magnetic fields that extended into the corona. Noise storms associated with spots with a south-seeking magnetic pole exhibited right-handed polarisation, and left-handed polarisation was found with north-seeking magnetic poles. The authors showed that these noise storms arose in the corona. In the case of the six outbursts, initially their

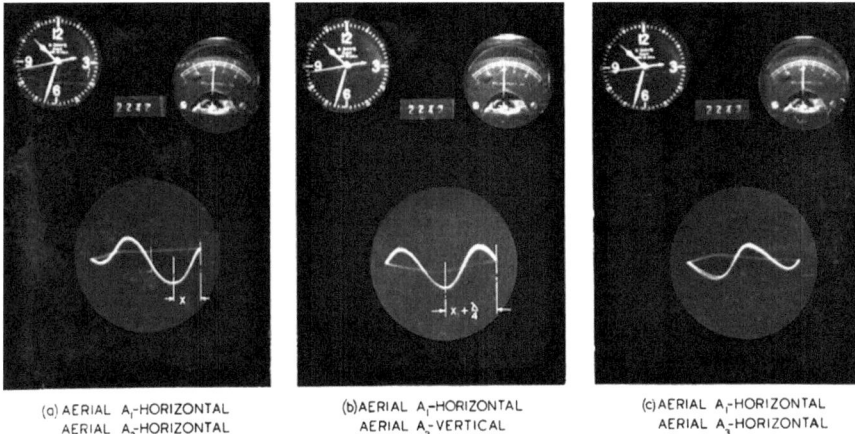

(a) AERIAL A₁-HORIZONTAL   (b)AERIAL A₁-HORIZONTAL   (c)AERIAL A₁-HORIZONTAL
   AERIAL A₂-HORIZONTAL   AERIAL A₂-VERTICAL   AERIAL A₃-HORIZONTAL

Fig. 3.14 Polarisation observations of a 97 MHz solar noise storm made with the Potts Hill interferometer (after Little and Payne-Scott, 1951: Plate 4)

positions and those of the related solar flares almost coincided, but the bursts were observed to move rapidly out from the Sun, at velocities of between 500 and 3000 km/s. Payne-Scott and Little suggested that the corpuscular streams responsible for terrestrial magnetic storms initiated the outbursts. These outbursts also showed the same polarisation features as observed with the noise storms.

Two exceptionally large outbursts occurred on 17 February and 21–22 February 1950, providing further evidence of the association of these rare events with solar flares and their terrestrial effects. Both outbursts lasted more than an hour and were observed from Potts Hill at 62, 97, 600, 1200, 3000 and 9400 MHz and from Mount Stromlo at 200 MHz, and reported on in a multi-authored paper published in 1951 in the *Australian Journal of Scientific Research*. Readings at four of these frequencies showed the two outbursts at times were strongly circularly polarised, while observations by Payne-Scott and Little with the two-element 97 MHz interferometer revealed that the sources of the emission began near the sites of the Hα flares, but moved progressively with the passage of time. In the case of the 17 February outburst, the emission began successively later as one moved from high to low frequencies, whereas the second outburst did not exhibit this tendency. In the case of both outbursts, the maximum radio output tended to occur earlier at higher frequencies.

Rod Davies (1930–2015) followed up this study with another that reviewed bursts detected between January 1950 and June 1951 at Potts Hill and Mount Stromlo, at the above-mentioned seven frequencies. He also looked at the connection between bursts, solar flares and terrestrial phenomena. Davies found that bursts (of all types) were far more prevalent at the lower frequencies – below 200 MHz – but that at higher frequencies a greater proportion of bursts tended to be of longer duration. He also suggested that "… there may be two separate components of bursts, one which shows rapid fluctuations and predominates at the lower frequencies, and one which is smooth and is characteristic of the high frequencies (although it may occur at low frequencies also)." (Davies, 1954: 90). He further suggested that these components may be due to plasma oscillations and thermal emission, respectively. He also found evidence of the correlations between bursts and both sunspots and flares noted by other researchers, but "… a burst on a high frequency is more likely to be accompanied by a flare than one on a low frequency." (Davies, 1954: 83). Short-wave radio fadeouts and magnetic crochets were overwhelmingly associated with bursts. Davies' paper was published in *Monthly Notices of the Royal Astronomical Society* after he had left RP and joined the Jodrell Bank radio astronomers in England, and it contains a variety of interesting statistics relating to terrestrial phenomena and solar bursts. For his recollections of the Potts Hill observations see Davies (2005, 2009).

3.1.3 Classifying solar bursts at Penrith and Dapto

Another way of researching bursts and outbursts was to look at their spectral features, and from 1948 Paul Wild was doing this with the new radiospectrograph at the Penrith field station (Fig. 2.45) (see Stewart et al., 2010). This novel radio telescope provided a virtually simultaneous visual display of radio emission over the frequency band 70–130 MHz. In the first instance, observations were made for a total of 254 hours between February and June 1949, and whenever a solar burst of interest was detected the trace of this was photographed on a cathode ray tube. Successive photographs could be taken at intervals of one-third of a second, permitting the radio astronomers to investigate the ways in which burst intensity changed with frequency and with time. From these photographs, Wild could construct the spectra of different bursts (Fig. 3.15). Producing these spectra manually was a very trying and time-consuming process – today it would all be done automatically by computer! Wild (1978) recalled that

> … in the early days we had a loaded camera, but we didn't actually have a motor so we had to turn it by hand. And we changed the aerial every twenty minutes … by … adjusting a rope, and Bill [Rowe] and I used to take twenty minute watches, just looking at the screen, waiting for something to happen. Of course we didn't know quite what to expect … but when the first few [bursts] came there was tremendous excitement.

With the benefit of hindsight, RP Chief, E.G. 'Taffy' Bowen (1911–1991) came to view the spectral study of solar bursts as one of the most important early achievements of the Radiophysics Lab (Bowen, 1973).

Analysis of the spectra of the bursts that were recorded during the first half of 1949 showed that many belonged to one of three distinct types, and these were designated Types I, II and III and described in a series of four papers written by Paul Wild in the *Australian Journal of Scientific Research* in 1950–51. Type I bursts (Fig. 3.16), described as "… little things popping up all over the place, like a choppy sea …" (Wild, 1978), occurred in large numbers (hundreds or more typically thousands) during so-called 'noise storms', which usually lasted for many hours, or even

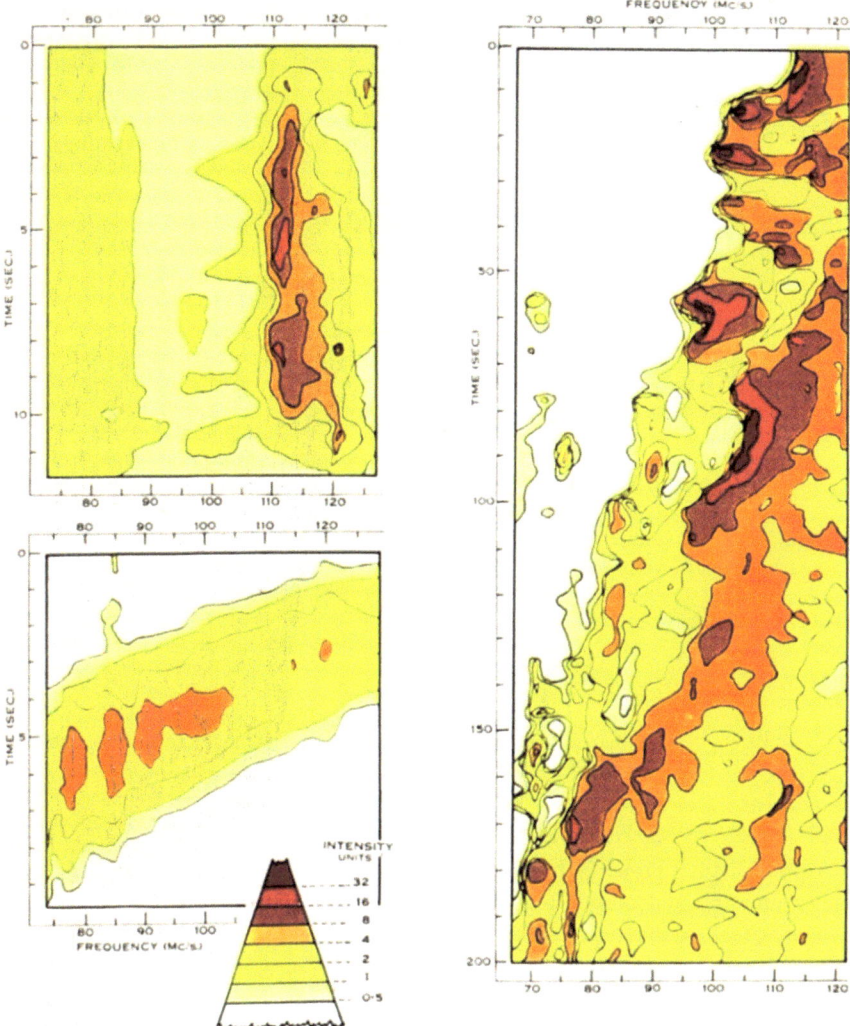

Fig. 3.15 Spectra of three quite different types of solar events constructed from data contained in the Penrith spectra (after Wild and McCready, 1950: Plate 2): Type I (upper left), Type II (right) and Type III (lower left)

days. Bursts normally came in small discrete groups, were strictly localised in both frequency (most were between 3 and 5 MHz) and time (typically 1–8 s), and showed strong circular polarisation. Type II bursts were rare, and outbursts like the ones shown in Fig. 3.15 were shown to be Type II events. These lasted from ten to tens of minutes, and had clearly-defined upper and lower frequency boundaries at any one point in time. The emission drifted from higher to lower frequencies with the passage of time at a mean rate of 0.22 MHz per second. Type II bursts often were associated with solar flares.

Time (minutes)

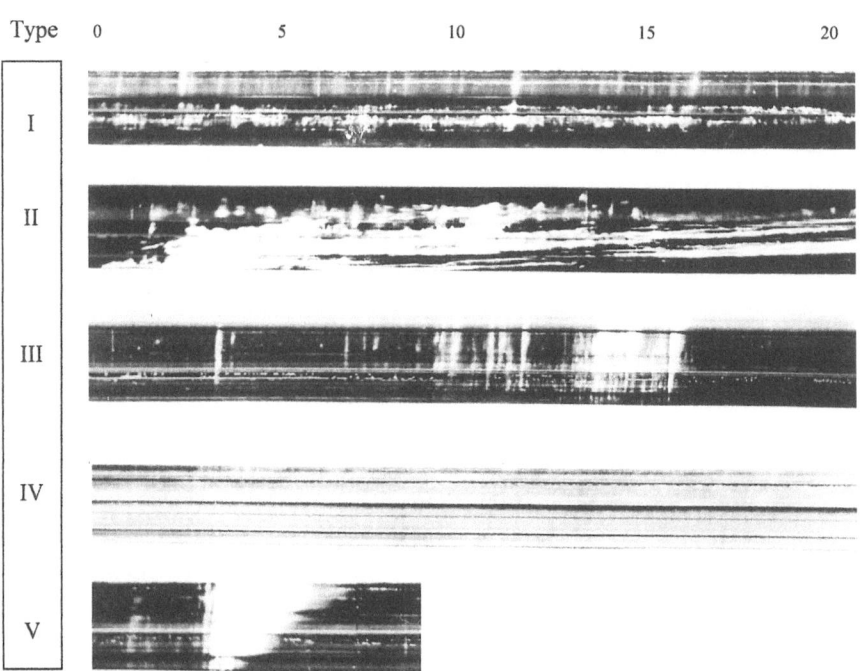

Fig. 3.16 Examples of the five main types of solar bursts taken with the Dapto spectrograph over the frequency range 15–200 MHz (courtesy: CRAIA)

A third distinct group of bursts belonged to Type III, characterised by narrow-band events that only lasted a few seconds and drifted rapidly from high to low frequencies (at mean rates of 20 MHz per second). Type III bursts were particularly common, and sometimes occurred in groups near the start of solar flares. Wild (1978) would later reminisce that at first he and his colleagues saw the identification of these three spectral types as "… a low key thing …", and it was only when Pawsey announced them at the Rome IAU meeting, causing "… a great deal of interest … that one realised that there was something to it."

Another interesting feature noted of some Type II and Type III bursts was that they sometimes exhibited harmonic structure, with a near mirror image of the initial burst following in close succession at a frequency separation of 2:1. The first Type II burst with harmonic structure was observed on 21 November 1952, just four months after the start of regular solar monitoring at Dapto, and was reported by Wild and assistants John Murray and Bill Rowe in research papers published in *Nature* and the *Australian Journal of Physics* in 1953 and 1954 respectively.

Once the Dapto field station was operational further observations were made, but over a wider frequency range (initially from 40 to 210 MHz), and actual spectra – such as those shown in Fig. 3.16 – were recorded on cathode ray tubes and were filmed directly with cine-cameras. This represented a quantum jump in efficiency; no longer was it necessary to generate spectra manually.

By 1958, three further spectral classes of solar events had been identified, which were respectively termed Type IV noise storms, Type V bursts and 'reverse drift pairs'. Type IV noise storms, well-documented at Dapto, but first described by a French radio astronomer, were rare continuum events, characterised by a high-intensity broadband featureless spectrum and linear polarisation. They lasted from around half an hour to six hours, and generally occurred after Type II bursts (but not *all* Type II bursts). Some of Payne-Scott's 'enhanced radiation' can be identified as Type IV events. Type V bursts looked like Type IIIs but with broadband continuum 'tails' that lasted anywhere from half a minute to three minutes and were associated with between 25% and 33% of all Type III bursts. Reverse drift pairs (RDPs) were first described by RP's Jim Roberts in 1958 (see Biobox 3.4). These rare very short-duration bursts were seen only at low frequencies (below 50 MHz), and occurred in pairs separated in time by only 1.5–2 s. The pairs typically drifted rapidly from lower to higher frequencies at rates of 2–8 MHz per second. RDPs tended to occur in storms lasting from hours to days, and about 10% were associated with weak Type III bursts. Roberts believed that RDPs were generated high in the solar corona.

Fig. 3.18 conveniently brings all of these spectral types of solar bursts together in the one diagram, and it is important to note that this RP classification scheme was quickly adopted worldwide.

During the 1950s as further spectral, positional and polarisation data on solar bursts emerged from Dapto, Paul Wild (Fig. 2.46) and his Solar Group colleagues began a series of detailed studies of the different types of events. This resulted in the

---

Biobox 3.4: Jim Roberts

James Alfred Roberts (Fig. 3.17) was born in Tamworth, New South Wales, on 27 April 1927, and completed BSc (First Class Honours) and MSc degrees at the University of Sydney in 1948 and 1949 respectively. He then undertook a PhD on solar bursts at Cambridge, supervised by the renowned English astronomer Fred Hoyle. From November 1952 he worked at the Radiophysics Lab initially as a member of the solar group. Jim then took up a postdoctoral fellowship (1958–61) at Caltech where he helped former RP colleagues John Bolton and Gordon Stanley establish the Owens Valley Radio Observatory in northern California.

Following his return to Australia, Roberts joined the group at Parkes and played a significant part in the commissioning of the new 64 m telescope. Although he took a lead role in observing radio emission from Jupiter, Jim considered himself a theoretical physicist first and a radio astronomer second. His research focus shifted to studies of the emission mechanisms from Galactic objects such as pulsars and from extragalactic objects such as quasars. He became well known for his review papers surveying particular branches of radio astronomy, such as his masterful 'Radio Emission from the Planets' published in 1963.

(continued)

Biobox 3.4 (continued)

Fig. 3.17 Jim Roberts began his career in solar radio astronomy and then transitioned to theoretical work on Galactic and extragalactic sources (courtesy: CRAIA)

At various times between 1973 and 1987 Roberts served as acting leader of the Astrophysics Group. He was also responsible for vetting all RP publications before their submission to journals, a role once played by Joe Pawsey. Jim took early retirement in 1987, by which time he had reached the rank of Senior Principal Research Scientist.

publication of a number of seminal papers. Don McLean (b. 1938; Fig. 3.19) found that Type I noise storms were relatively uncommon: between 1952 and 1958 only ten were recorded at Dapto. These often followed, and presumably were associated with, solar flares. The available evidence suggested that some sources giving rise to Type I emission moved upwards through the solar corona with velocities of ~100 km/s, while others showed downwards motion. Wild (1957: 324) noted that "Such behaviour is reminiscent of ascending and descending prominence material." In 1958 Max Komesaroff (d. 1988; Fig. 3.19) confirmed the earlier findings of Payne-Scott and Little that Type I noise storms were strongly polarised. By the early 1960s, there was a unique assemblage of Dapto spectral, positional and polarisation data at RP, and in 1964 Alan Weiss and Wayne Orchiston (b. 1943; see Fig. 3.2) drew on this to prepare a detailed review of Type I storms, but at the end of that year Weiss died from leukemia and this paper was never published. Consequently, by 1965 less was known about Type I storms than any other spectral type.

In a long review paper published in 1959, Jim Roberts summarised the findings from a study of 65 different Type II bursts observed at Dapto between September 1952 and March 1958. In another, shorter paper he noted that Type II bursts were rare: "Even at sunspot maximum the average rate of occurrence is only about one burst every 50 hours, so that a long series of observations is needed to define the characteristics of the bursts." (Roberts, 1959: 194). About 60% of all Type II bursts

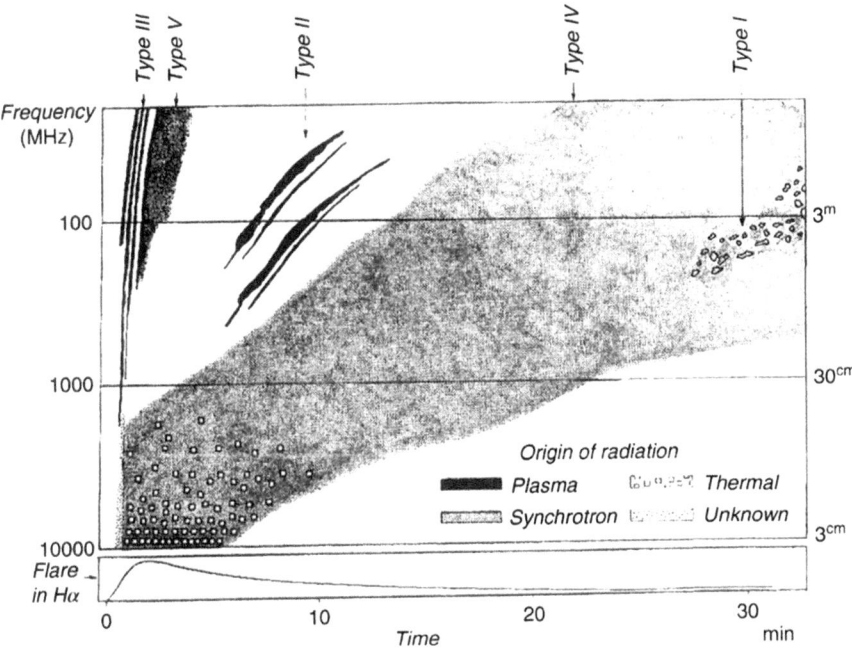

Fig. 3.18 The five radio burst types following a solar flare, with the origin of the mechanism producing the radiation (courtesy: CRAIA)

Fig. 3.19 Two members of the Dapto group were Don McLean (left) and Max Komesaroff. Don showed that Type I noise storms are relatively uncommon, while Max confirmed earlier evidence that Type I storms are strongly polarised (courtesy: CRAIA)

were preceded some minutes earlier by a group of Type III bursts. The optical solar expert Ronald Giovanelli (1915–1984) from CSIRO's Division of Physics and Roberts found that most Type II bursts were also associated with optical features, such as flares, surges, ejecting prominences or disappearing filaments. Quantitative studies revealed that in about 60% of cases, Type II bursts exhibited fundamental and second harmonic bands, and in about 20% of these that the two bands were themselves split and typically separated by 4–8 MHz. Stefan ('Steve') Smerd (1916–1978) and colleagues at RP noticed an interesting feature of Type II bursts exhibiting harmonic structure: the harmonic radiation always originated from a lower region of the corona than the fundamental emission.

On the basis of positional observations, Weiss reported that some Type II events were in fact made up of two or more bursts that occurred simultaneously or in quick succession, with source positions and rates of frequency drift that differed widely. Most Type II bursts were featureless, but around 20% of them exhibited considerable fine structure. One of the most prominent forms was the so-called 'herring-bone structure', characterised by elements that drift rapidly towards higher and lower frequencies. The first example of herring-bone structure was recorded at Dapto in April 1956. Type II bursts are thought to be caused by shock fronts that cause plasma oscillations as clouds of ionised matter pass through successively higher regions of the corona with velocities of 750 to 1500 km/s; these values are derived from Weiss' analysis of observations made with the position interferometer at Dapto. It is likely that these same streams of particles are the ones that impact the Earth 1–3 days after a flare, and cause geomagnetic disturbances (including aurorae). Where Type II bursts are associated with Type III bursts, the evidence very strongly suggests "… that the sources of the emission of these two distinctive types of burst are located in equivalent regions in the corona … [although] the exciters of these two types of burst differ widely in physical nature and in speed of motion through the corona …" (Weiss, 1963a: 263).

Unlike Type II bursts, Type III bursts are common, occurring at a rate of about one every few minutes. In a study of over 300 flares and microflares made between November 1955 and July 1956, Giovanelli's optical colleague from CSIRO's Division of Physics, Ralph Loughhead (1929–2018), together with Jim Roberts and Marie McCabe from the RP Lab, showed that about 20% were associated with Type III bursts, with the figure rising to about 60% for flares of Class 1 or greater. Ron Giovanelli found an even higher correlation between Type III bursts and flares that exhibited what he called 'puffs', sudden explosive expansion that occurred at the outbreak of some flares.

Contrary to the earlier findings of Komesaroff, when he analysed about 100 Type III bursts captured by the Dapto swept-frequency interferometer between 1960 and 1962, visiting Indian scientist Gopala Rao (Fig. 3.2) found that the great majority were either only slightly or moderately polarised. Just 13% of all bursts showed polarisation readings in excess of 40%, and the degree of polarisation for all bursts varied inversely as the duration of the burst. Positional observations of Type III bursts made in 1958 and reported in two 1959 papers by Wild and three RP colleagues revealed that radiation at decreasing frequencies originated from successively higher levels in the corona (Fig. 3.20), confirming the impression that the emission was due to plasma oscillations caused by disturbances that moved rapidly out from the Sun at velocities of 0.2–0.8 times the speed of light. These rates are considerably greater than those inferred earlier from spectral data alone, and a subsequent study of 50 Type III bursts by Ron Stewart (b. 1939; Fig. 3.21) produced a mean figure of 0.33. However, the positional observations showed clearly that the corona was considerably denser in regions where Type III bursts were generated than in other parts of the corona, suggesting that these bursts were associated with

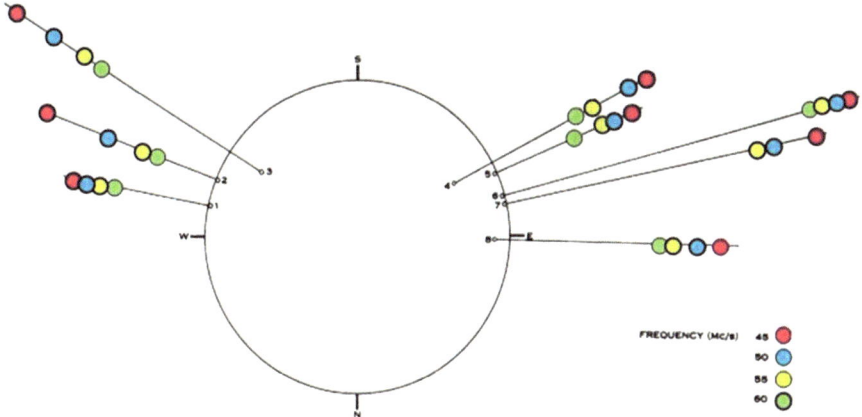

Fig. 3.20 Positions of Type III bursts observed at different frequencies (adapted from Wild et al., 1959a: 382)

Fig. 3.21 A study of Type III bursts by Ron Stewart showed that they move out rapidly from the Sun, on average, at one-third the speed of light. After his retirement, Ron completed a PhD on the history of the Penrith and Dapto field stations and their contribution to international radio astronomy (courtesy: CRAIA)

coronal streamers. The disturbances giving rise to the Type III bursts may correlate with the cosmic ray showers that reach the Earth about one hour after a major flare.

Alan Weiss published a major review paper on Type IV storms in 1963, based on observations carried out at Dapto over the frequency range 40–70 MHz, between 1952 and 1961. During this period 17 Type IV storms were recorded, highlighting their comparative rarity. Weiss (1963b: 530) found that "The sources of type IV bursts almost invariably exhibit movement at some stage during their lifetimes. The movement may be irregular, or consist of quasi-periodic oscillations about some stable position, or be systematic in time." Even when there was clear evidence of motion, this was generally small and the average height of Type IV sources was ~2.11 solar radii, with little or no difference in position at different frequencies in the 45–65 MHz range. This indicates that the emission cannot result from plasma oscillations, and synchrotron radiation (electrons spiraling in a magnetic field) is the most likely explanation. Weiss termed these 'stationary' Type IV events, and distinguished them from the much rarer 'moving' Type IV events, which typically were of shorter duration and were associated with a source that moved rapidly out through the solar corona. Weiss believed it is highly likely that the two different varieties of Type IV event "… are distinct phenomena related only through a common initiating disturbance low in the solar atmosphere." (Weiss, 1963b: 541).

Indian visitor, Thiruvenkata Krishnan (1934–2019, Fig. 3.22), and RP's Dick Mullaly (1924–2001; Fig. 2.5) observed a number of Type IV storms with the 1420 MHz Chris Cross at Fleurs and found that at this frequency these events originated in the chromosphere or lower corona, and did not exhibit any obvious motion outwards from the Sun. The sizes of the emitting regions were also small at this frequency, measuring just 2–5 arcsec in diameter. Don McLean noted that Type IV storms almost always were associated with major solar flares and that they were followed by geomagnetic storms. With the advent of the Culgoora Radioheliograph and real-time television-like images of solar events, Paul Wild and his colleagues would come to realise that their earlier ideas about Type IV events were over-simplified. This sophisticated new radio telescope gave a completely new picture of Type IV bursts and showed that they were not one type but rather a range of different phenomena.

Type V bursts were first described by Wild, Kevin Sheridan (Biobox 3.5) and Gil Trent, and by Alan Neylan (d. 2015), in 1959 on the basis of spectral observations at 40–240 MHz made at Dapto in 1957 and 1958. Wild, Sheridan and Neylan (1959a: 393) pointed out that "In some cases the emission appears merely as a diffuse prolongation of the type III burst, in others as a detached blob or patchiness." In a sample of 27 Type V bursts, Neylan found that 20 of them (i.e. 74%) were associated with bursts that occurred nearly simultaneously at between 1000 and 9400 MHz (whereas the great majority of 'standard' Type III bursts (96.2%) were unaccompanied by emission at these higher frequencies), so this association with higher frequency emission can be regarded as a diagnostic feature of these bursts. But as with Type III events, most Type

Fig. 3.22 Indian solar astronomer Thiruvenkata Krishnan (right) with Chris Christiansen at the Fleurs field station (courtesy: Krishnan collection; all rights reserved)

V bursts were associated with flares, particularly what Giovanelli (1958) termed 'flare puffs'. Several Type V bursts investigated by Wild, Sheridan and Trent (1959b: 182) exhibited outward motion, at velocities ~3000 km/s, and they noted that "… while the duration, intensity, and frequency range of these enhancements differ markedly from those of the type IV bursts, their transverse motion appears to be similar." Alan Weiss and Ron Stewart later revised this figure downwards to ~2000 km/s. Moreover, "… at the start of the burst the type V source does not, in general, coincide in position with the associated type III source." (Weiss and Stewart, 1965: 147). There is no evidence of harmonic structure and Type V bursts are, at most, weakly polarised. In explaining the origin of a Type V burst, Weiss and Stewart tentatively suggested that part of the cloud of electrons that generates the associated Type III burst is trapped in the corona, and it is these electrons which cause the plasma oscillations that produce the Type V emission. Much harder to explain are the rarer detached Type V bursts.

Sometimes when data from the Dapto radiospectrographs, the position interferometer and the solar polarimeter were combined they showed that what at first sight might seem to be simple solar events were in fact exceptionally complex. For example, a Type III event might be followed in quick succession by a Type V burst, a Type II outburst and a Type IV storm. By pooling the various Dapto data, the RP

Biobox 3.5: Kevin Sheridan
Kevin V. Sheridan (Fig. 3.23) was born in Brisbane on 4 August 1918 and died in Sydney in 2010. He received a BA in mathematics from the University of Sydney, and BSc and DSc degrees from the University of Queensland. In 1945 he joined the Radiophysics Lab and after working on aircraft distance measuring equipment, Joe Pawsey invited him to join the radio astronomy group, where he became involved in the design, construction and early use of the Mills Cross at Fleurs.

Fig. 3.23 Kevin Sheridan specialised in the design and development of instruments for solar radio astronomy (courtesy: CRAIA)

(continued)

Biobox 3.5 (continued)

Sheridan then became a leading member of the solar group, and played a key role in the development of the Dapto radiospectrographs and swept-frequency interferometer at the Dapto field station. Kevin's contribution to the design and development of the Culgoora Radioheliograph in northern NSW was central to the success of the instrument. In a review of the Radioheliograph, Paul Wild wrote: "I should acknowledge in particular the instrumental work of Kevin Sheridan (without whom the thing would never have worked)." (Wild, 1974: 3).

From 1978 until his retirement in 1983 Sheridan was head of the RP solar group. In 1966 he was appointed the Foundation Secretary of the newly-formed Astronomical Society of Australia. Kevin was made a Member of the Order of Australia (AM) in 1984 for his significant service to radio astronomy.

radio astronomers were able to investigate the spectral signatures of these events, and temporal variations in source position and size, in polarisation and in total intensity of the emission.

3.1.4   Studies of the 'quiet Sun' at Potts Hill

When single-frequency observations were carried out at frequencies well above the initial upper limit of the Dapto radiospectrograph (i.e. 210 MHz), few solar bursts were found to be present, the notable exceptions being Type II outbursts and some Type IV storms. Instead, observations during the early 1950s with Christiansen's grating arrays at the Potts Hill field station revealed the existence of localised regions of enhanced emission, which were located in the chromosphere and lower corona and survived for one or sometimes two or even three solar rotations. Spectrohelioscopic observations carried out concurrently at Potts Hill showed that these radio-active regions usually were situated above large sunspot groups and chromospheric hydrogen and calcium plage regions, and so they were assigned the name 'radio plages'. The daily motion of these radio plages reflected the rotation of the Sun (e.g. see Fig. 3.24). We now know that these radio plages are responsible for the slowly-varying component of solar radiation.

Chris Christiansen and Don Mathewson (b. 1929; Fig. 3.25) presented the 'latest word' on these radio plages at the 1958 Paris Symposium on Radio Astronomy and in a multi-authored paper published in the *Annales d'Astrophysique* (Christiansen et al., 1960). But it was only with the construction of the innovative Chris Cross at Fleurs that it became possible to obtain precise two-dimensional data on these features and confirm their chromospheric associations (e.g. see Fig. 3.26). From 1957, the advent of daily 1420 MHz solar maps allowed the Fleurs radio astronomers to

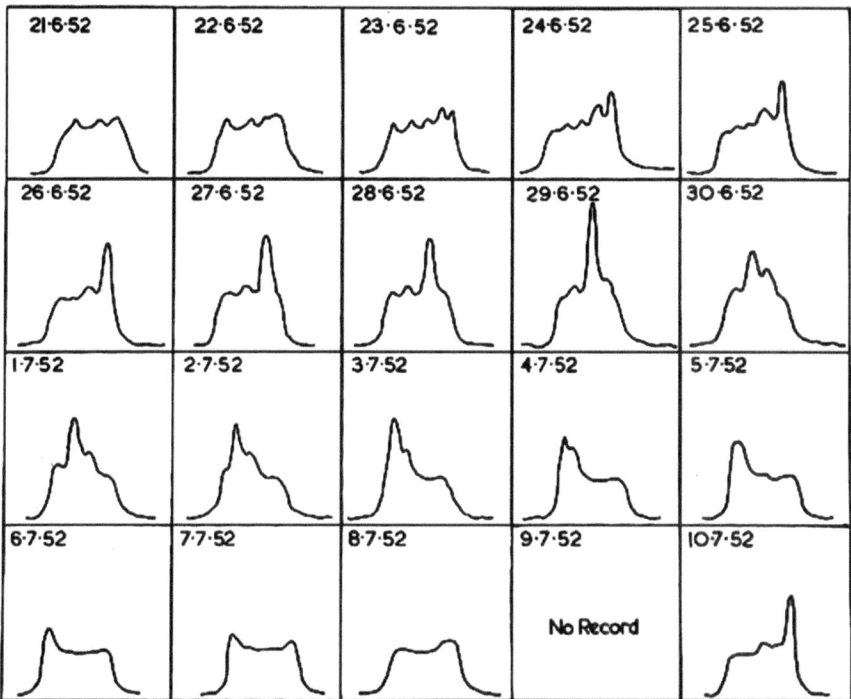

Fig. 3.24 Averaged daily east–west scans of the Sun showing the development and decay of radio plages (courtesy: CRAIA)

Fig. 3.25 Don Mathewson began his career in solar radio astronomy and then in 1958 he began a PhD at Jodrell Bank using the newly completed 250 ft dish. In 1966 he joined the staff at the Mt Stromlo Observatory and became one of the very first radio astronomers to make the transition to optical astronomy. He served as Director of the Observatory from 1977 to 1986 (courtesy: Mt Stromlo and Siding Spring Observatories)

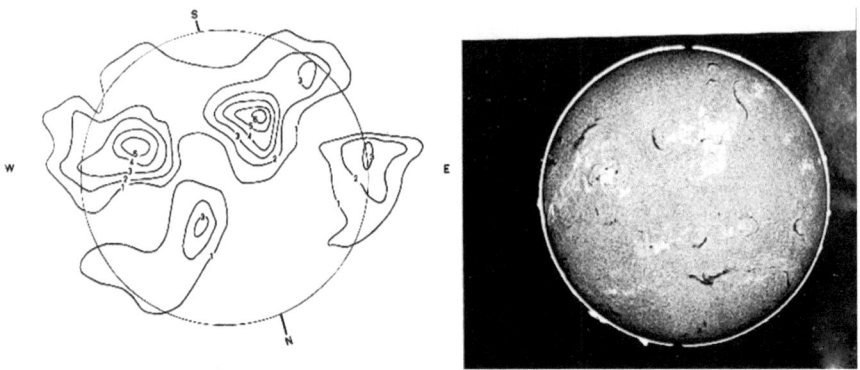

Fig. 3.26 Isophote map of solar radio emission at 1420 MHz, showing its correlation with chromospheric features (courtesy: CRAIA)

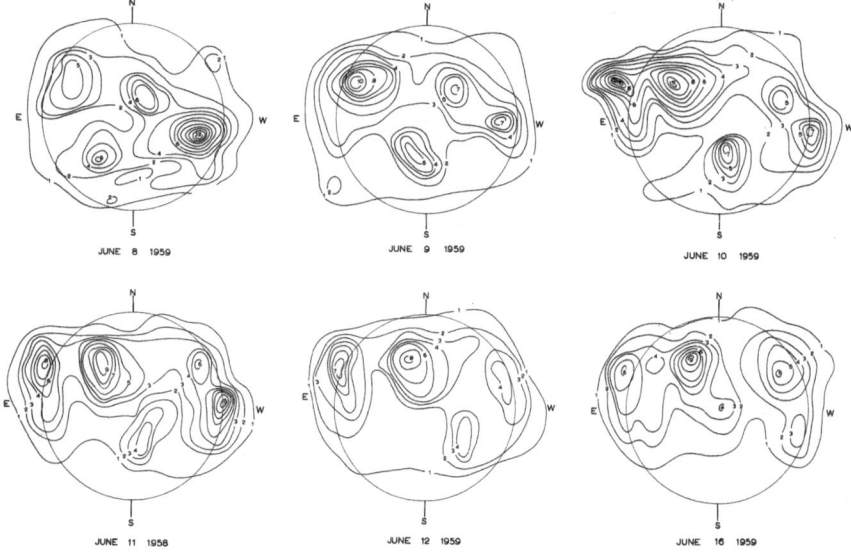

Fig. 3.27 A succession of 1420 MHz solar maps obtained at Fleurs in June 1959 (courtesy: CRAIA)

follow the evolution of these radio plages over the course of a single solar rotation. Fig. 3.27 shows a succession of such solar maps, as published in the IAU's *Quarterly Bulletin on Solar Activity*. Radio plages have typical diameters of 2–6 arcmin (representing 100,000–300,000 km in actual areal extent), are situated from 30,000 to 100,000 km above the photosphere (with an average height of ~40,000 km), and have peak temperatures of less than 200,000 K up to about 1,600,000 K (with a median value of ~600,000 K). It would appear that these radio plages physically "…

consist of large clouds of gas (principally hydrogen) ... [that are] much denser than the surrounding atmosphere ... [and] are prevented from dissipating presumably by magnetic fields." (Christiansen and Mullaly, 1963: 171). The virtual absence of circular polarisation indicates that the emission from these radio plages is thermal in origin.

Not all RP solar researchers were interested in burst emission or active regions. Some were interested in determining the nature of the 'quiet Sun' – its quiescent level when burst emission and radio-active regions were absent. Harry Minnett and Norman Labrum (Fig. 3.28) were able to reflect on this back in 1948 when they analysed data from the 1 November solar eclipse. They concluded that the solar radiation received at 9400 MHz could be explained in terms of two quite different models: (1) that 74% of the radio emission came from the Sun's visible disk, with the balance from a bright ring around the circumference; or (2) all radiation came from a uniform disk with a diameter 1.1 times that of the optical Sun.

The first alternative suggested that at 9400 MHz the Sun exhibited 'limb-brightening', in marked contrast to the distinctive 'limb-darkening' observed at optical wavelengths, so this proposition was easy to test. Chris Christiansen and Joe Warburton (1923–2005) began by superimposing numbers of east–west scans obtained with the 32-element solar grating array at Potts Hill over a nine month period (Fig. 3.29). By ignoring all active regions and isolating the common lower envelopes of these accumulated curves, they were able to derive a mean profile of the 'quiet Sun' which supported the limb-brightening proposition (Christiansen and Warburton, 1953a, 1953b). Their results were consistent with those developed earlier by their RP colleague, Steve Smerd (Biobox 3.6), on the basis of theoretical considerations.

With the advent of the second (16-element) Potts Hill grating array, Christiansen and Warburton (1955) were able to synthesise 'processed' north–south and east–west scans obtained between 1952 and 1954 and derive "... a two-dimensional brightness distribution of the "quiet" Sun ..." at 1420 MHz (see also Christiansen et al., 1957). The resulting radio isophote plot (Fig. 3.31) shows clear pictorial

Fig. 3.28 Norm Labrum had a long career at RP working mostly on solar radio emission, culminating in the comprehensive and highly-regarded book *Solar Radiophysics*, which he co-edited with Don McLean (see McLean and Labrum, 1985) (courtesy: CRAIA)

Fig. 3.29 Superimposed east–west scans of the Sun at 1420 MHz obtained with the original Potts Hill grating array in 1952 (after Christiansen and Warburton, 1953a: 200)

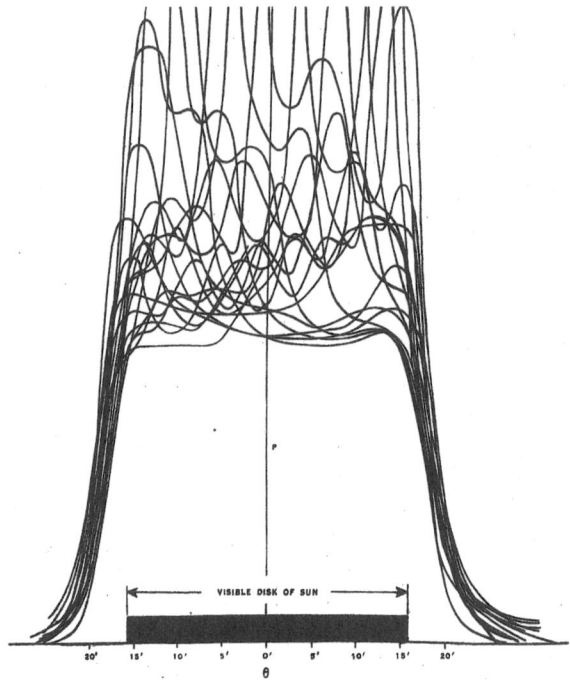

Biobox 3.6: Steve Smerd

Stefan Frederick Smerd (Fig. 3.30) was born in Austria in 1916. He studied physics at the Technische Hochschule in Vienna before fleeing from the Nazis in 1938 and moving to England. In 1942 he completed a BSc at the University of Liverpool, and spent the remainder of WWII working in the microwave magnetron laboratory at the University of Birmingham and at the Admiralty Signals Establishment.

Fig. 3.30 Steve Smerd was the resident theorist in the RP solar group (courtesy: CRAIA)

(continued)

Biobox 3.6 (continued)

In 1946 Smerd joined the Radiophysics Lab, first working in the valve group, and then in early 1947 he became a member of the solar group with responsibility for theoretical issues relating to the solar observations. Steve rapidly gained an international reputation for his work on thermal processes in the 'quiet Sun', and for his research on the extremely energetic non-thermal phenomena associated with solar disturbances. Joe Pawsey referred to Steve as a 'walking encyclopaedia' on the subject. Smerd went on to establish a world data centre for solar radio emission during the International Geophysical Year (1957–58) and he became an active member of the International Astronomical Union.

In 1971 Smerd succeeded his long-term colleague Paul Wild as head of the solar group and Director of the solar observatory at Culgoora. Steve and his guitar were always welcome at RP parties and other functions (see Fig. 2.51). He died in December 1978 during heart surgery. For an entertaining biography see Wild (1980); see also Orchiston (2014b) and Robertson (2002).

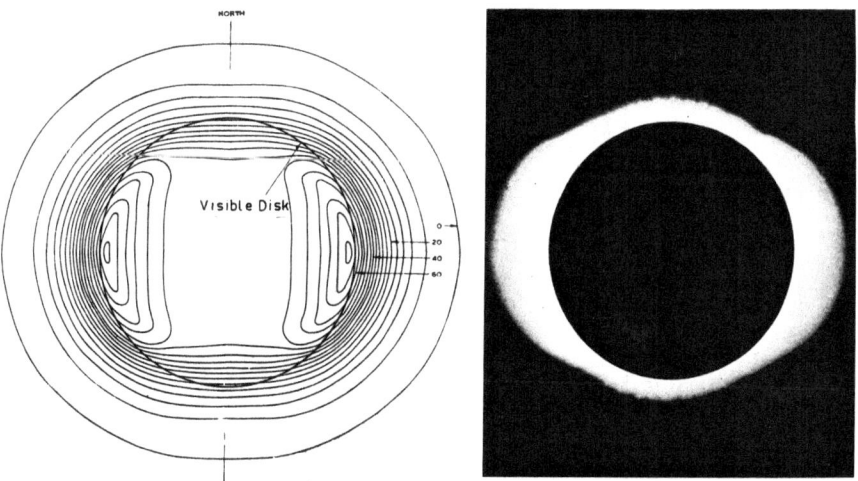

Fig. 3.31 Isophote map of the 'quiet Sun' at 1420 MHz in 1952–53, together with a photograph of the solar corona (after Christiansen and Warburton, 1955: 482 and Plate 2)

evidence of limb-brightening. Moreover, the radio Sun is seen to be non-circular, confirming the earlier suspicions of both optical and radio researchers, with the limb-brightening confined to the near-equatorial regions. Christiansen and Warburton cautioned that these results were obtained from observations taken at or near sunspot minimum; given that the 1420 MHz emission derived in the main from the corona, it was reasonable to assume that the temporal changes in the brightness distribution would occur in the course of a sunspot cycle as the form and extent of the (optical) corona underwent change. Nevertheless, Christiansen regarded the 1955 paper as particularly important given that "[We] got this earth-rotational synthesis put down [in print]. And it was also very important as … it hadn't been suspected that it [limb brightening] would only be a purely equatorial effect … This was quite a new sort of thing." (Christiansen, 1976).

In 1958, near sunspot maximum, Norm Labrum used the Chris Cross to test Christiansen and Warburton's proposition that the solar brightness distribution at 1420 MHz should change in the course of a solar cycle. Data derived from north–south strip scans of the Sun and from daily isophote maps both produced similar results: an apparent disk temperature of ~140,000 K, or twice the value obtained at sunspot minimum. Even more surprising, Labrum (1960: 700) found that "There is limb darkening at the poles, and the distribution does not appear to have changed in shape between sunspot minimum (1953) and the time of the present series of observations." Visiting Indian radio astronomer T. Krishnan and Labrum (1961) confirmed this conclusion in a later paper, where they combined data obtained using a radiometer and the Chris Cross during the partial solar eclipse of 8 April 1959. Further information about solar research carried out using the Chris Cross is provided by Orchiston and Mathewson (2009).

One final aspect of RP's solar research program should be mentioned. Between 1957 and 1960, Bruce Slee used the north–south arm of the Mills Cross (Fig. 2.59) and a number of antennas at remote sites to investigate the outer solar corona and the solar wind. Interference fringes associated with 13 discrete sources were examined as they passed close to the Sun with a view to establishing the extent of coronal scattering. He found that "Sporadic large increases in the scattering first became noticeable when the angular separation was as much as $100R_\odot$ and at separations of less than $60R_\odot$ the effects of scattering could be detected on every record." (Slee, 1961: 225). These findings are illustrated graphically in Fig. 3.32.

Two RP staff, Jack Piddington and Steve Smerd, were primarily theoreticians, and used their knowledge and mathematical skills to investigate the nature of the Sun's atmosphere. Piddington was particularly proud of the first paper he published in this area, which appeared in the prestigious *Proceedings of the Royal Society* in 1950. Titled, 'The derivation of a model solar chromosphere from radio data', this was

> … the first model that used the radio data … With the other models being developed [by other researchers] if you tested them with the radio data you immediately found that they failed dismally … [My model] caused a moderate stir in those days amongst the theorists." (Piddington, 1978).

In reflecting on the success of RP's solar work and its pre-eminent status in world radio astronomy during the 1940s through to the 1970s, Steve Smerd remarked:

> Wild built up a group which was quite unusual compared to all the other research teams I have seen. I think our Solar Group under Paul was perhaps the happiest, frictionless collection of people you can imagine. We were keen and dedicated and would have done anything that Paul even half-mentioned or suggested, let alone explicitly asked. (Smerd, 1978).

This beautifully summarises the ethos of the RP Solar Group in the 1960s, when one of the authors of this book (WO) enjoyed several years as a junior member of its ranks.

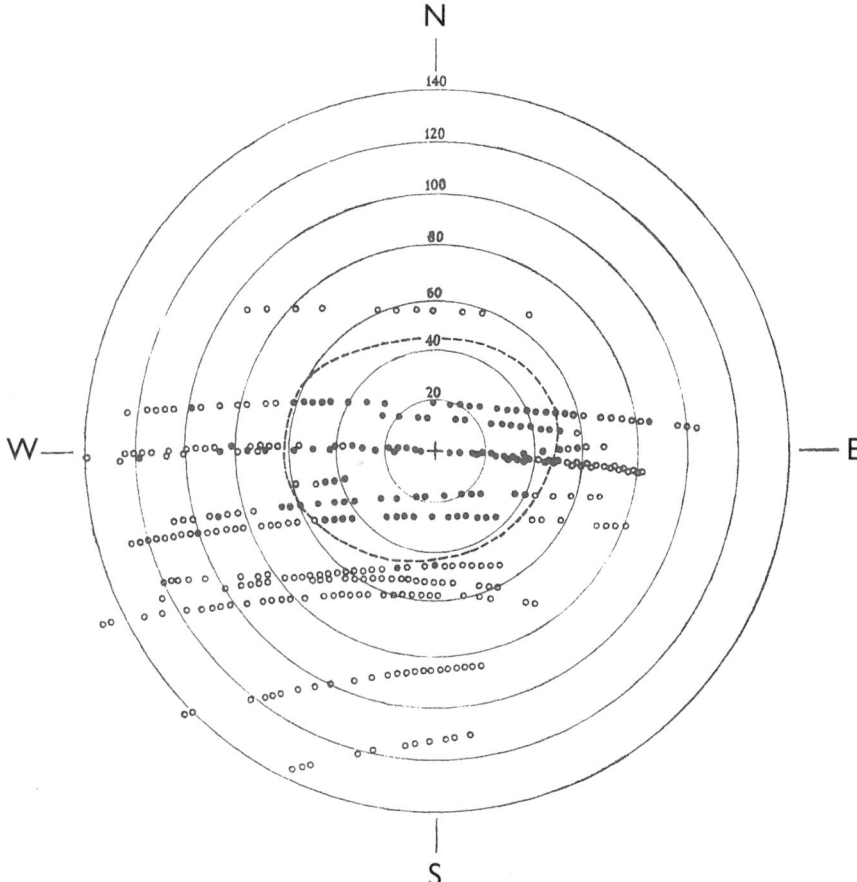

Fig. 3.32 Evidence of coronal scattering is indicated by the closed circles, and is primarily restricted to the region enclosed by the dashed line. The Sun is marked by the cross at centre, and the concentric circles are at 20 solar radii intervals (after Slee, 1961: 227)

3.2   The Moon

The Sun was not the only body in our Solar System to interest the RP radio astronomers. Back in the 1950s, the Moon and Jupiter were also known to emit radio waves, and both were targets of research.

Three different lunar astronomy programs were carried out by RP staff during the field stations era, although only two of these were associated with field stations. All three projects were short-lived, and two took place in the year 1948 and were probably inspired by the appearance of a paper by Robert H. Dicke (1916–1997) and Robert Beringer in the *Astrophysical Journal* (Dicke and Beringer, 1946). In this paper the two Americans reported observing the Moon at a frequency of 24 GHz, and deriving a disk temperature of 292 K. Two years later, Jack Piddington

(Biobox 3.7) and Harry Minnett (Biobox 3.8) noted that lunar temperature varied according to phase, and that Dicke and Beringer's value therefore required a correction factor, which produced a corrected figure of 270 K.

Piddington and Minnett then decided to investigate the Dicke and Beringer result empirically, and between April and July 1948 they carried out observations at an identical frequency with a small radio telescope mounted on the 'Eagle's Nest', atop the Radiophysics Lab (Fig. 2.9). They expected that the radio emission would directly mimic the lunar phase, but to their surprise it did not. Years later Minnett (1978) was to reflect on this: "It was a strange thing and that got us excited straight away. It had this phase lag of three days." They ended up making observations over three lunar cycles, which yielded a temperature of 234 K for subsurface layers of the Moon's crust. These results were consistent with the existence of a thin layer of dust covering the solid lunar

---

Biobox 3.7: Jack Piddington

Jack Hobart Piddington (Fig. 3.33) was born in Wagga Wagga, New South Wales, on 6 November 1910 and died in Sydney on 16 July 1997. He completed a combined BE–BSc degree (for which he was awarded the University Medal), an MSc degree at the University of Sydney, and in 1938 a PhD on ionospheric work, under Edward Appleton, at Cambridge. Upon joining the Radiophysics Lab in 1939, he worked on the wartime development of radar – including units suitable for tropical climates (see Fig. 1.7).

After the War Piddington worked initially on distance measuring equipment for civilian aviation before transferring to the radio astronomy group. Over the next two decades he made valuable contributions to observational

Fig. 3.33 Jack Piddington was best known for his work on theoretical problems in radio astronomy (courtesy: CRAIA)

(continued)

Biobox 3.7 (continued)

radio astronomy, including the discovery with Harry Minnett of the intense source Sagittarius A at the centre of our Galaxy (see next chapter). However, Jack was best known for his theoretical contributions to our understanding of a range of astronomical phenomena. In 1967 he transferred to the CSIRO Division of Physics and was promoted to Chief Research Scientist.

Piddington received the Syme Medal from the University of Melbourne in 1958 and the T.K. Sidey Medal from the Royal Society of New Zealand the following year. He was a Visiting Professor at the Universities of Maryland (1960) and Iowa (1967) and was elected a Fellow of the Royal Astronomical Society. Apart from his numerous research papers, Jack was also known for his books, *Radio Astronomy* (1961) and *Cosmic Electrodynamics* (1969, 1981). For further details see Melrose and Minnett (1998) and Orchiston (2014c).

Biobox 3.8: Harry Minnett

Harry Clive Minnett (Fig. 3.34) was born in Hurstville, Sydney on 12 June 1917 and died in Sydney on 20 December 2003. After completing a BSc–BE double degree at the University of Sydney in 1939, he accepted a post at the newly-formed Radiophysics Lab, where he worked on wartime radar developments. After the war he carried out lunar and solar research in collaboration with Jack Piddington. They were the first to detect the intense radio source Sagittarius A, shown later to be the centre of our Galaxy. Harry also worked on radio navigational aids for civilian aircraft and a traffic radar system for the police force (Fig. 1.10).

Fig. 3.34 In 1978 Harry Minnett became the fifth Chief of Radiophysics, following on from Paul Wild (courtesy: CRAIA)

(continued)

Biobox 3.8 (continued)

Following the decision in 1954 to construct the Parkes Radio Telescope, Minnett spent four years in London liaising with the firm of Freeman Fox on the planning and design of the instrument. Once it was built he was responsible for surveying the surface of the dish and overseeing its subsequent upgrades. He also spent two years in Canberra on secondment as project manager for the design and construction of the optical 3.9 m Anglo-Australian Telescope. After returning to Radiophysics, Harry worked on the Interscan Microwave Landing System, and from 1978 to 1981 he was Chief of the Division. Following his retirement, Harry continued to take an interest in Radiophysics and its successor, the Australia Telescope National Facility, proudly presenting a paper on 'Fifty Years of Radio Science and its Applications' at the IAU General Assembly in Sydney in 2003. For further details see Orchiston (2014d) and Thomas and Robinson (2005).

surface, and this interpretation was confirmed in person by the Apollo 11 astronauts 21 years later. In September 1950, two years after their initial study, Piddington and Minnett recorded lunar emission at 1210 and 3000 MHz from Potts Hill, obtaining values of 212 ± 64 K and 215 ± 65 K respectively for the apparent disk temperature.

In a very different project, Frank Kerr (1918–2000) and Alex Shain (1922–1960; Fig. 2.28) decided to bounce radar signals off the Moon in order to investigate properties of the Earth's upper atmosphere. Like Piddington and Minnett, they too were responding to earlier American experiments, reported in the April 1946 issue of *Electronics*, in which engineers from the US Army Signal Corps had successfully recorded radar signals bounced off the Moon. However, the Americans found large variations in signal strength with time, presumably of ionospheric origin, and it was this feature that attracted Kerr and Shain. Their experiment was carried out in 1948–49 using the signals broadcast at 21.54 and 17.84 MHz from a 'Radio Australia' station at Shepparton, Victoria, and received at RP's Hornsby Valley field station. Thirty different experiments were carried out over a year, and echoes were received on 24 occasions (Fig. 3.35). From our viewpoint, the interesting part of their reports, published in *Nature* (Kerr et al., 1949) and subsequently in *Proceedings of the Institute of Radio Engineers* (Kerr and Shain, 1951), is the conclusion that they drew regarding the nature of the lunar surface. Kerr and Shain found the radar echoes showed two types of fading, one due to ionospheric effects, and the other as a result of the Moon's libration. This fact, together with the elongation of short pulses on reflection, showed that the Moon's rocky surface was 'rough' rather than smooth on scales of tens of metres.

After undertaking the Moon-bounce experiment, Kerr (1971) spent a year "… studying and writing up all the Moon work and also thinking about possibilities of echoes from the Sun and other planets." He published his calculations and ideas about doing radar far beyond the Moon in the *Proceedings of the Institute of Radio Engineers* (Kerr, 1952), and since it was the first paper ever written on this topic he later came to regard it as a classic. RP scientists, however, decided not to do any

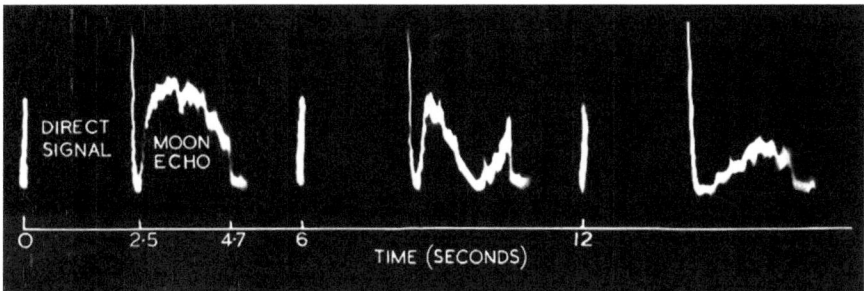

Fig. 3.35   Three successive Moon echoes, showing rapid fading due to libration (the slow rotation of the Moon on its axis) (after Kerr and Shain, 1951: 232)

further radar astronomy "... because it was quite clear that it would involve building a pretty expensive transmitter and antenna system." (Kerr, 1971). Funding for major projects like this would always be a problem within the limited RP budget.

3.3   Jupiter

In 1955 American radio astronomers, Bernie Burke (1928–2018) and Ken Franklin (1923–2007), surprised the astronomical world by reporting the existence of Jovian bursts at the low frequency of 22 MHz (Franklin, 1984). Although RP scientist Alex Shain and colleague Charlie Higgins were quick to report confirmatory observations, this serendipitous discovery of radio emission from Jupiter was one of RP's most notable 'lost opportunities'. Back in 1950–51, Shain and Higgins were at Hornsby Valley investigating Galactic emission at 18.3 MHz, and on various occasions they recorded intervals of intense static (Fig. 3.36) which they attributed to terrestrial interference – or so it seemed at the time. Immediately following the Burke and Franklin announcement, Shain revisited the 1950–51 records and discovered that some of their episodes of 'interference' occurred when Jupiter was within the beam of their radio telescope. Furthermore, when the occurrence of these bursts was plotted against Jupiter's rotation period (of just under 10 hours) he noticed that they were not uniformly distributed in Jovian longitude but instead clustered between 0° and 135° (see Fig. 3.37). In other words, the radiation seemed to come from a localised region of the planet with a period of rotation of 9h 55m 13 ± 5s, remarkably close to the so-called System II value (Fig. 3.38) based on visual observations.

Shain was intrigued by these bursts. Where did they originate? He began by examining the atmospheric features of Jupiter, and noticed that when the Jovian bursts were recorded in 1950–51 there was a group of visual spots associated with the South Temperate Belt that also rotated in 9h 55m 13s – precisely the same as the radio source. Could these be the cause of the radio bursts, and if so, what kind of

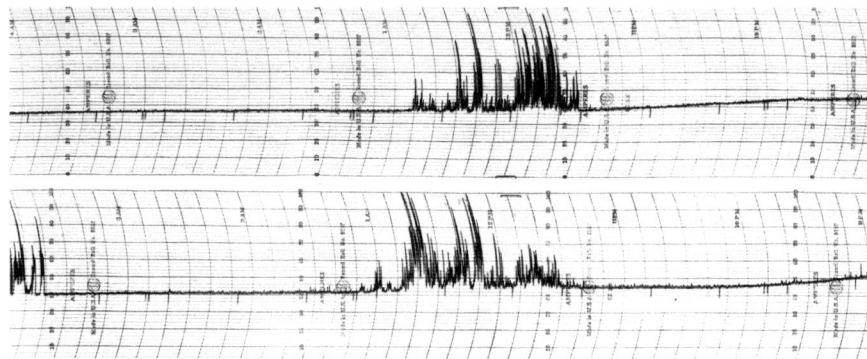

Fig. 3.36 Examples of 18.3 MHz Jovian bursts recorded on 17 October 1950 (above) and 29 October 1950 (below) by Shain and Higgins, five years before Burke and Franklin's reported discovery (after Shain, 1956: 65)

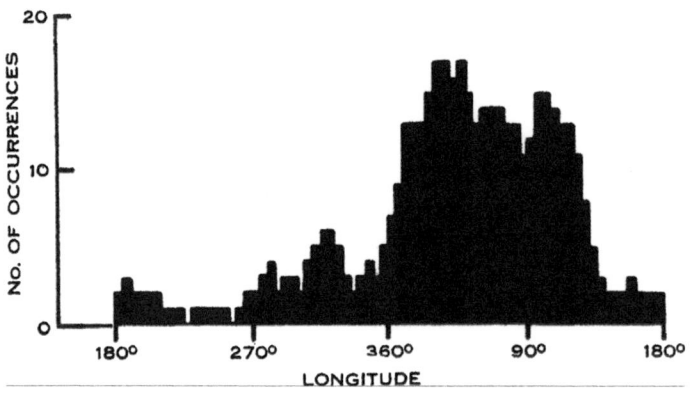

Fig. 3.37 Occurrence frequency of 18.3 MHz Jovian emission for 5° intervals of longitude and assuming a rotation period of 9h 55m 13s (after Shain, 1956: 68)

emission mechanism was involved? In order to pursue this interesting research challenge further, Shain and Frank Gardner (Biobox 3.9) observed Jupiter between June 1955 and March 1956, using a 19.6 MHz interferometer at Fleurs, and occasionally the newly-constructed east–west arm of the 19.7 MHz Shain Cross (Fig. 2.63). Between November 1955 and March 1956 they also used antennas operating at 14 and 27 MHz at Fleurs, and a small 19.7 MHz antenna located at Potts Hill.

Jovian emission was regularly recorded at 19.6 MHz, generally in the form of groups of bursts that varied rapidly in intensity, with large-scale changes taking place over time intervals as short as 0.2 s. Similar variations characterised bursts observed at 27 MHz, but unlike the 19.6 MHz bursts – which were comparatively common – these were only detected on about 20% of all observing nights. The results at 14 MHz were severely affected by manmade interference, but Jovian bursts also appeared to be far less frequent than at 19.6 MHz. Meanwhile, from

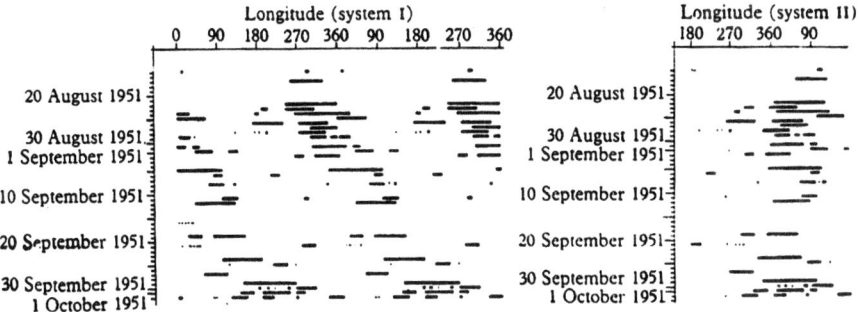

Fig. 3.38 Occurrence of 18.3 MHz Jovian emission plotted against the longitude of the central meridian at the time of the observations. The sloping bars in the left-hand plot indicate the rotation period of the Jovian radio source differs markedly from that of System I, whereas the near-vertical stacking of the bars in the right-hand plot shows it is very similar to System II (after Shain, 1956: 67)

observations at all three frequencies, Gardner and Shain (1958) concluded that peak burst intensity probably occurred at around 20 MHz.

Biobox 3.9: Frank Gardner

Francis Frederick Gardner (Fig. 3.39) was born in Sydney in 1924. He obtained degrees in Science and Electrical Engineering (First Class Honours) from the University of Sydney in 1943 and 1945, before completing a PhD on ionospheric properties at Cambridge University in 1949. He joined the Radiophysics Lab in 1950 and worked on ionospheric research until 1955 when he began investigating Jovian decametric emission at Fleurs and Potts Hill in collaboration with Alex Shain.

Fig. 3.39 Frank Gardner made a major contribution to the list of research successes during the early days of the Parkes Radio Telescope (courtesy: CRAIA)

(continued)

Biobox 3.9 (continued)

In 1957 Frank (known affectionately as 'FF') began developing low-noise amplifiers for the Parkes Radio Telescope. From 1962 until his retirement in 1989 he carried out research at Parkes that produced cutting edge results in fields as diverse as the polarisation of Galactic radio emission and interstellar chemistry through molecular line studies. Frank was also a key member of the Parkes team that carried out extensive sky surveys first at 408 MHz and then at 2.7 GHz.

Two of his close colleagues wrote: "Frank was always modest, quiet and unassuming. He was endowed with dry humour and a sense of fun that made working with him anything but boring. These characteristics were enhanced by an uncanny 'feel' for microwave engineering …" (Milne and Whiteoak, 2005). FF died in 2002 at the age of 78.

These findings merely built on those Shain came up with in his analysis of the 1950–51 observations, and when he and Gardner examined the longitude distribution of the 19.6 MHz bursts they found a rotation period for the source of 9h 55m 34s, close to the earlier result (Fig. 3.40). However, when they considered the 19.6 and 27 MHz data separately, these tended to indicate the presence of three different sources: the main one at a Jovian longitude in System II of 0° and two much less active secondary sources at longitudes of –100° and +80°. Furthermore, the apparent correlation between the main source and visual spots in the South Temperate Belt – as suggested by the 1950–51 evidence – was not sustained. In the light of these latest observations, and others made by Burke and Franklin, Shain and Gardner considered the emission mechanism responsible for the bursts and concluded that they were generated by plasma oscillations associated with an ionised region in the atmosphere of Jupiter.

Alex Shain's intention was to conduct simultaneous radio and optical observations of Jupiter, with a view to investigating any optical features that could conceivably be associated with this emission mechanism, but his untimely death in 1960 put paid to these plans. Instead, it was Bruce Slee and Charlie Higgins who took up the challenge, and during Jupiter's 1962 opposition they employed long baseline interferometry in order to investigate the identity and nature of the emitting source (or sources). For this project, they used square arrays of 19.7 MHz half-wave dipoles, one located at Fleurs and the other at a remote site, Freeman's Reach, about 32 km to the north. Their observations showed that the "… angular diameter of a burst source was … less than a third of the planet's diameter, and that all bursts contributing to a noise storm originated in a single area less than a half of the size of the planet's disk." (Slee and Higgins, 1963: 781–82).

Slee and Higgins decided to move to longer baselines, and in 1963 and 1964 they employed effective spacings between 10 and 200 km. Simple arrays of dipoles at Fleurs and three other sites, Dapto and Jamberoo south of Sydney, and Heaton to the north (see Fig. 2.8), were phased and oriented to receive Jovian emission over an

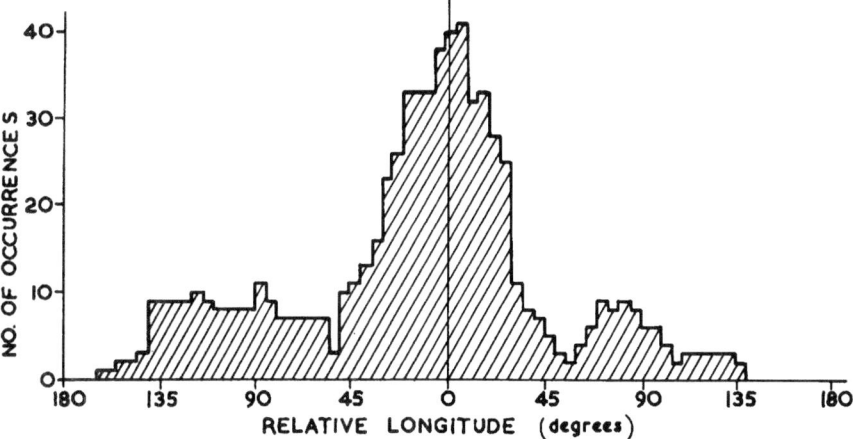

Fig. 3.40 Occurrence frequency of 19.6 MHz Jovian emission for 5° intervals of longitude and assuming a rotation period of 9h 55m 34s (after Gardner and Shain, 1958: 66)

hour-angle range of ±4 hr, and a radio link was used to transmit the signals from the remote sites to Fleurs.

Analysis of the observations suggested that the emitting regions were typically 10–15 arcsec in size, but Slee and Higgins concluded that they were probably very much smaller and that scattering in the interplanetary medium gave anomalously large results. This conclusion turned the Jovian decametric project at RP in a new direction: what had started as a quest for emission source size now became an investigation of scattering by the interplanetary medium. Slee and Higgins then used their 1963–64 data on burst arrival times, burst rates, angular position scintillations and apparent angular size to successfully investigate interplanetary diffraction patterns and electron irregularities in the solar wind.

This would mark the final contribution of the Fleurs site to planetary radio astronomy. We now know that the Jovian bursts are associated with spiraling electrons in the magnetic torus that extends from the inner moon, Io, to Jupiter's magnetosphere. They have nothing whatsoever to do with the spots or other features seen in Jupiter's ever-changing atmospheric 'cloud belts'.

References

Allen, C.W., 1947. Solar radio-noise at 200 Mc./s. and its relation to solar observations. *Monthly Notices of the Royal Astronomical Society*, 107, 386–396.

Bowen, E.G., 1947. Letter to S.B. Nicholson, dated 21 January. In CSIRO Archives, file A1/3/1a. Sullivan collection.

Bowen, E.G., 1973. Taped interview with Woody Sullivan, dated 24 December.

Christiansen, W.N., 1976. Taped interview with Woody Sullivan, dated 27 August.

Christiansen, W.N., and Mullaly, R.F., 1963. Solar observations at a wavelength of 20 cm with a crossed-grating interferometer. *Proceedings of the Institution of Radio Engineers Australia*, 24, 165–173.

Christiansen, W.N., and Warburton, J.A., 1953a. The distribution of radio brightness over the solar disk at a wavelength of 21 cm. Part I. A new highly directional aerial system. *Australian Journal of Physics*, 6, 190–202.

Christiansen, W.N., and Warburton, J.A., 1953b. The distribution of radio brightness over the solar disk at a wavelength of 21 cm. Part II. The quiet Sun – one-dimensional observations. *Australian Journal of Physics*, 6, 262–271.

Christiansen, W.N., and Warburton, J.A., 1955. The distribution of radio brightness over the solar disk at a wavelength of 21 cm. Part III. The quiet Sun – two-dimensional observations. *Australian Journal of Physics*, 8, 474–486.

Christiansen, W.N., Warburton, J.A., and Davies, R.D., 1957. The distribution of radio brightness over the solar disk at a wavelength of 21 cm. Part IV. The slowly varying component. *Australian Journal of Physics*, 10, 491–514.

Christiansen, W.N., Mathewson, D.S., Pawsey, J.L., et al., 1960. A study of a solar active region using combined optical and radio techniques. *Annales d'Astrophysique*, 23, 75–101.

Christiansen, W.N., Yabsley, D.E., and Mills, B.Y., 1949. Measurements of solar radiation at a wavelength of 50 centimetres during the eclipse of November 1, 1948. *Australian Journal of Scientific Research*, A2, 506–523.

Davies, R.D., 1954. An analysis of bursts of solar radio emission and their association with solar and terrestrial phenomena. *Monthly Notices of the Royal Astronomical Society*, 114, 74–92.

Davies, R.D., 2005. A history of the Potts Hill radio astronomy field station. *Journal of Astronomical History and Heritage*, 8, 87–96.

Davies, R.D., 2009. Recollections of two and a half years with 'Chris' Christiansen. *Journal of Astronomical History and Heritage*, 12, 4–10.

Débarbat, S., Lequeux, J., and Orchiston, W., 2007. Highlighting the history of French radio astronomy. 1: Nordmann's attempt to observe solar radio emission in 1901. *Journal of Astronomical History and Heritage*, 10, 3–10.

Denisse, J.-F., 1949. Relation entre les émissions radioélectriques solaires décimétriques et les taches du soleil. *Comptes Rendus de l'Academie des Sciences*, 228, 1571–1572.

Dicke, R.H., and Beringer, R., 1946. Microwave radiation from the Sun and Moon. *Astrophysical Journal*, 103, 375.

Franklin, K.L., 1984. The discovery of Jupiter bursts. In Kellermann, K.I., and Sheets, B. (eds.). *Serendipitous Discoveries in Radio Astronomy*. Green Bank (WV), National Radio Astronomy Observatory. Pp. 252–257.

Gardner, F.F., and Shain, C.A., 1958. Further observations of radio emission from the planet Jupiter. *Australian Journal of Physics*, 11, 55–69.

Giovanelli, R.G., 1958. Flare-puffs as a cause of Type III radio bursts. *Australian Journal of Physics*, 11, 350–352.

Harris, M., 2017. *Rocks, Radio and Radar: The Extraordinary Scientific, Social and Military Life of Elizabeth Alexander*. New Jersey, World Scientific.

Haynes, R., Haynes, R., Malin, D., and McGee, R., 1996. *Explorers of the Southern Sky. A History of Australian Astronomy*. Cambridge, Cambridge University Press.

Hey, J.S., 1946. Solar radiations in the 4–6 metre radio wave-length band. *Nature*, 157, 47–48.

Kerr, F.J., 1952. On the possibility of obtaining radar echoes from the Sun and planets. *Proceedings of the Institute of Radio Engineers*, 40, 660–666.

Kerr, F.J., 1971. Taped interview with Woody Sullivan, dated 3 October.

Kerr, F.J., and Shain, C.A., 1951. Moon echoes and transmission through the ionosphere. *Proceedings of the Institute of Radio Engineers*, 39, 230–242.

Kerr, F.J., Shain, C.A., and Higgins, C.S., 1949. Moon echoes and penetration of the ionosphere. *Nature*, 163, 310–313.

Krishnan, T., and Labrum, N.R., 1961. The radio brightness distribution on the Sun at 21 cm from combined eclipse and pencil-beam observations. *Australian Journal of Physics*, 14, 403–419.

Labrum, N.R., 1960. The radio brightness of the quiet Sun at 21 cm wavelength near sunspot maximum. *Australian Journal of Physics*, 13, 700–711.

Lehany, F.J., and Yabsley, D.E., 1948. A solar noise outburst at 600 Mc./s and 1,200 Mc./s. *Nature*, 161, 645–646.

Lehany, F.J., and Yabsley, D.E., 1949. Solar radiation at 1200 Mc/s, 600 Mc/s and 200 Mc/s. *Australian Journal of Scientific Research*, A2, 48–62.

Little, A.G., and Payne-Scott, R., 1951. The position and movement on the solar disk of sources of radiation at a frequency of 97 Mc/s. *Australian Journal of Scientific Research*, A4, 489–507.

McCready, L.L., Pawsey, J.L., and Payne-Scott, R., 1947. Solar radiation at radio frequencies and its relation to sunspots. *Proceedings of the Royal Society*, A190, 357–375.

McLean, D.J., and Labrum, N.R. (eds), 1985. *Solar Radiophysics – Studies of Emission from the Sun at Metre Wavelengths*. Cambridge, Cambridge University Press.

McNally, D., 1990. Obituary. C.W. Allen (1904–1987). *Quarterly Journal of the Royal Astronomical Society*, 31, 259–266.

Martyn, D.F., 1946. Polarisation of solar radio-frequency emissions. *Nature*, 158, 308.

Melrose, D.B., and Minnett, H.C., 1998. Jack Hobart Piddington 1910–1997. *Historical Records of Australian Science*, 12, 239–246.

Mills, B.Y., 1976. Taped interview with Woody Sullivan, dated 25 August.

Milne, D.K., and Whiteoak, J.B., 2005. The impact of F.F. Gardner on our early research with the Parkes Radio Telescope. *Journal of Astronomical History and Heritage*, 8, 33–38.

Minnett, H.C., 1978. Taped interview with Woody Sullivan, dated 2 March.

Orchiston, W. 2004a. From the solar corona to clusters of galaxies: the radio astronomy of Bruce Slee. *Publications of the Astronomical Society of Australia*, 21, 23–71.

Orchiston, W., 2004b. The 1948 solar eclipse and the genesis of radio astronomy in Victoria. *Journal of Astronomical History and Heritage*, 7, 118–121.

Orchiston, W., 2005. Dr Elizabeth Alexander: first female radio astronomer. In Orchiston, W. (ed.). *The New Astronomy: Opening the Electromagnetic Window and Expanding Our View of Planet Earth*. Dordrecht, Springer. Pp. 71–92.

Orchiston, W., 2014a. Martyn, David Forbes. In Hockey, T. et al. (eds.). *Biographical Encyclopedia of Astronomers. Second Edition*. New York, Springer. Pp. 1405–1406.

Orchiston, W., 2014b. Smerd, Stefan Friedrich. In Hockey, T. et al. (eds.). *Biographical Encyclopedia of Astronomers. Second Edition*. New York, Springer. Pp. 2019–2020.

Orchiston, W., 2014c. Piddington, Jack Hobart. In Hockey, T. et al. (eds.). *Biographical Encyclopedia of Astronomers. Second Edition*. New York, Springer. Pp. 1714–1715.

Orchiston, W., 2014d. Minnett, Harry Clive. In Hockey, T. et al. (eds.). *Biographical Encyclopedia of Astronomers. Second Edition*. New York, Springer. Pp. 1495–1496.

Orchiston, W., 2016. Elizabeth Alexander and the mysterious 'Norfolk Island Effect'. In Orchiston, W. (ed.). *Exploring the History of New Zealand Astronomy: Trials, Tribulations, Telescopes and Transits*. Cham (Switzerland), Springer. Pp. 629–651.

Orchiston, W., and Mathewson, D., 2009. Chris Christiansen and the Chris Cross. *Journal of Astronomical History and Heritage*, 12, 11–32.

Orchiston, W., and Slee, B., 2002. The Australasian discovery of solar radio emission. *AAO Newsletter*, 101, 25–27.

Orchiston, W., Nakamura, T., and Strom, R. (eds), 2011. *Highlighting the History of Astronomy in the Asia–Pacific Region: Proceedings of the ICOA-6 Conference*. New York, Springer.

Orchiston, W., Sim, H., and Robertson, P., 2016. Dr Owen Bruce Slee, 10 August 1924 – 18 August 2016. *Australian Physics*, 53, 214.

Orchiston, W., Slee, B., and Burman, R., 2006. The genesis of solar radio astronomy in Australia. *Journal of Astronomical History and Heritage*, 9, 35–56.

Pawsey, J.L., 1946. Observation of million degree radiation from the Sun at a wavelength of 1.5 metres. *Nature*, 158, 633.

Pawsey, J.L., 1950. Solar radio-frequency emission. *Proceedings of the Institution of Electrical Engineers*, 97, 290–310.

Pawsey, J.L., Payne-Scott, R., and McCready, L.L., 1946. Radio-frequency energy from the Sun. *Nature*, 157, 158–159.

Payne-Scott, R., 1949. Bursts of solar radiation at metre wavelengths. *Australian Journal of Scientific Research*, A2, 214–227.

Payne-Scott, R., and Little, A.G., 1951. The position and movement on the solar disk of sources of radiation at a frequency of 97 Mc/s. *Australian Journal of Scientific Research*, A4, 508–525.

Payne-Scott, R., Yabsley, D.E., and Bolton, J.G., 1947. Relative times of arrival of bursts of solar noise at different radio frequencies. *Nature*, 160, 256–257.

Piddington, J.H., 1978. Taped interview with Woody Sullivan, dated 13 March.

Piddington, J.H., and Oliphant, M.L., 1972. David Forbes Martyn. *Records of the Australian Academy of Science*, 2(2), 47–60.

Reber, G., 1944. Cosmic static. *The Astrophysical Journal*, 100, 279–287.

Roberts, J., 1959. Some aspects of Type II bursts. In Bracewell, R. (ed.). *Paris Symposium on Radio Astronomy*. Stanford, Stanford University Press. Pp. 194–200.

Robertson, P., 2002. Stefan Friedrich Smerd (1916–1978): CSIRO solar physicist. In Ritchie, J. (ed.). *Australian Dictionary of Biography, Volume 16*. Melbourne, Melbourne University Press. P. 266.

Shain, C.A., 1956. 18.3 Mc/s radiation from Jupiter. *Australian Journal of Physics*, 9, 61–73.

Slee, O.B., 1961. Observations of the solar corona out to 100 solar radii. *Monthly Notices of the Royal Astronomical Society*, 123, 223–231.

Slee, O.B., and Higgins, C.S., 1963. Long baseline interferometry of Jovian decametric radio bursts. *Nature*, 197, 781–782.

Smerd, S., 1978. Taped interview with Woody Sullivan, dated 6 March.

Stewart, R., Wendt, H., Orchiston, W., and Slee, B., 2011a. A retrospective view of Australian solar radio astronomy 1945 to 1960. In Orchiston, W. et al., 2011. Pp. 589–629.

Stewart, R., Orchiston, W., and Slee, B., 2011b. The Sun has set on a brilliant mind: John Paul Wild (1923–2008), solar radio astronomer extraordinaire. In Orchiston, W. et al., 2011. Pp. 527–542.

Stewart, R., Wendt, H., Orchiston, W., and Slee, B., 2010. The Radiophysics field station at Penrith, New South Wales, and the world's first solar radio spectrograph. *Journal of Astronomical History and Heritage*, 13, 2–15.

Sullivan, W.T., 1982. *Classics in Radio Astronomy*. Cambridge, Cambridge University Press.

Sullivan, W.T., 2009. *Cosmic Noise: A History of Early Radio Astronomy*. Cambridge, Cambridge University Press.

Thomas, B.M., and Robinson, B.J., 2005. Harry Clive Minnett, 1917–2003. *Historical Records of Australian Science*, 16, 199–220.

Weiss, A.A., 1963a. The positions and movements of the sources of solar radio bursts of spectral Type II. *Australian Journal of Physics*, 16, 240–271.

Weiss, A.A., 1963b. The Type IV solar radio burst at metre wavelengths. *Australian Journal of Physics*, 16, 526–544.

Weiss, A.A., and Stewart, R.T., 1965. Solar radio bursts of spectral type V. *Australian Journal of Physics*, 18, 143–165.

Wendt, H., and Orchiston, W., 2018. The contribution of the AN/TPS-3 radar antenna to Australian radio astronomy. *Journal of Astronomical History and Heritage*, 21, 65–80.

Wendt, H., Orchiston, W., and Slee, B., 2008. The Australian solar eclipse expeditions of 1947 and 1949. *Journal of Astronomical History and Heritage*, 11, 71–78.

Wild, J.P., 1953. Techniques for observation of radio-frequency radiation from the Sun. In Kuiper, G.P. (ed.). *The Solar System. Volume 1. The Sun*. Chicago, University of Chicago Press. Pp. 676–692.

Wild, J.P., 1957. Spectral observations of solar activity at metre wavelengths. In Hulst, H.C. van de (ed.). *Radio Astronomy*. Cambridge, Cambridge University Press. Pp. 321–326.

Wild, J.P., 1974. A new look at the Sun. In Contopoulus, G. (ed.). *Highlights of Astronomy*, Volume 3. Dordrecht, Reidel. Pp. 3–19.

Wild, J.P., 1978. Taped interview with Woody Sullivan, dated 3 March.

Wild, J.P., 1980. The Sun of Stefan Smerd. In Kundu, M.R., and Gergely, T.E. (eds). *Radio Physics of the Sun*. Dordrecht, Reidel. Pp. 5–21.

Wild, J.P., and McCready, L.L., 1950. Observations of the spectrum of high-intensity solar radiation at metre wavelengths. I. The apparatus and spectral types of solar bursts observed. *Australian Journal of Scientific Research*, A3, 387–396.

Wild, J.P., Sheridan, K.V., and Neylan, A.A., 1959a. An investigation of the speed of the solar disturbances responsible for Type III radio bursts. *Australian Journal of Physics*, 12, 369–398.

Wild, J.P., Sheridan, K.V., and Trent, G.H., 1959b. The transverse motions of the sources of solar radio bursts. In Bracewell, R. (ed.). *Paris Symposium on Radio Astronomy*. Stanford, Stanford University Press. Pp. 176–185.

Chapter 4
Expanding Horizons – The Milky Way and Beyond

4.1 Discovery of the First Discrete Radio Sources

Non-solar research at the Radiophysics Laboratory (RP) was launched in September 1946 when Joe Pawsey (1908–1962) tried unsuccessfully to observe the enigmatic 'radio star' in Cygnus that the British team of Stanley Hey (1909–2000, Fig. 4.1), John Parsons and James Phillips had announced in the 17 August 1946 issue of *Nature* (Hey et al., 1946). Two months later, John Bolton (1922–1993; Fig. 4.2) and research assistant Bruce Slee (1924–2016) were at Dover Heights trying to observe the Sun at 60 MHz. When it insisted on remaining inactive they decided to use their 2-Yagi antenna in sea interferometer mode to search for radio emission from other types of objects. Neither had a background in astronomy, and their astronomical knowledge and resources were virtually non-existent. Bolton (1982: 349) later described how they used the Russell, Duggan and Stewart book *Astronomy* to "… hazard guesses as to which types of objects might emit copious amounts of radio emission …" and *Norton's Star Atlas* "… to find the position of the brightest candidate in each class."

Their search was both unsuccessful and unauthorised, and on this latter count attracted the ire of their boss, Joe Pawsey (Fig. 4.3): "Our efforts … were cut short by an unheralded visit from Pawsey, who noted that the aerials were not looking at the Sun. Suffice it to say that he was not amused and we were both ordered back to the Lab." (Bolton, 1982: 349–350).

By early 1947 Bolton had done ample penance and was allowed to return to Dover Heights, but this time with Pawsey's blessing he searched for the Cygnus source and other radio stars whenever the Sun was inactive. By June it was, and Bolton teamed with receiver specialist Gordon Stanley (1921–2001) to search for the Cygnus source with the 100 MHz sea interferometer. They easily detected the source (Fig. 4.4), and saw ample evidence of the anomalous intensity variations reported by Stanley Hey and his group. Meanwhile, the interference fringes showed that this radio star was less than 8 arcmin in diameter, the first evidence for the very

© The Author(s) 2021
W. Orchiston et al., *Golden Years of Australian Radio Astronomy*,
Historical & Cultural Astronomy, https://doi.org/10.1007/978-3-319-91843-3_4

Fig. 4.1 Stanley Hey (1909–2000) pioneered radio astronomy in Britain. After World War II, Hey and his group reported intense variable radio emission from the Cygnus constellation (courtesy: Royal Radar Establishment; all rights reserved)

Fig. 4.2 John Bolton on the roof of the Dover Heights blockhouse in May 1947, shortly before the detection of the Cygnus source. The two elements of the 100 MHz Yagi were converted into a sea interferometer and used to discover the first eight discrete sources (courtesy: Stanley family)

Fig. 4.3 John Bolton
(left), Gordon Stanley and
Joe Pawsey in one of the
RP instrument rooms
[courtesy: CSIRO Radio
Astronomy Image Archive
(CRAIA)]

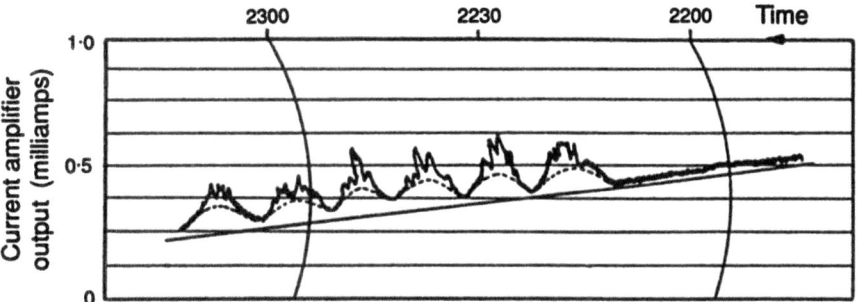

Fig. 4.4 The interference pattern from Cygnus A recorded over a one-hour period on the evening of 19 June 1947 at Dover Heights, showing the characteristic spiky structure in the signal. The dashed curve provides a measure of the angular size of the source (after Bolton and Stanley, 1948a: 60)

small angular size of the source. Subsequently, daily rising patterns with the antenna pointed in various north–east directions yielded a very approximate position. Over the next two to three months the Cygnus source was also detected with the 60, 85 and 200 MHz Yagis and a rough radio spectrum was plotted (subsequently found to be inaccurate). Their exciting results were reported in issues of both *Nature* (Bolton and Stanley, 1948b) and the *Australian Journal of Scientific Research* (Bolton and Stanley, 1948a; Fig. 4.5). In 1950 Jack Piddington (1910–1997; Fig. 3.33) and Harry Minnett (1917–2003; Fig. 3.34) extended the spectrum to 1210 MHz, and confirmed the non-thermal nature of the emission.

In November 1947 Bruce Slee joined Bolton and Stanley, and the three young researchers used aerials at Dover Heights and at Long Reef and West Head,

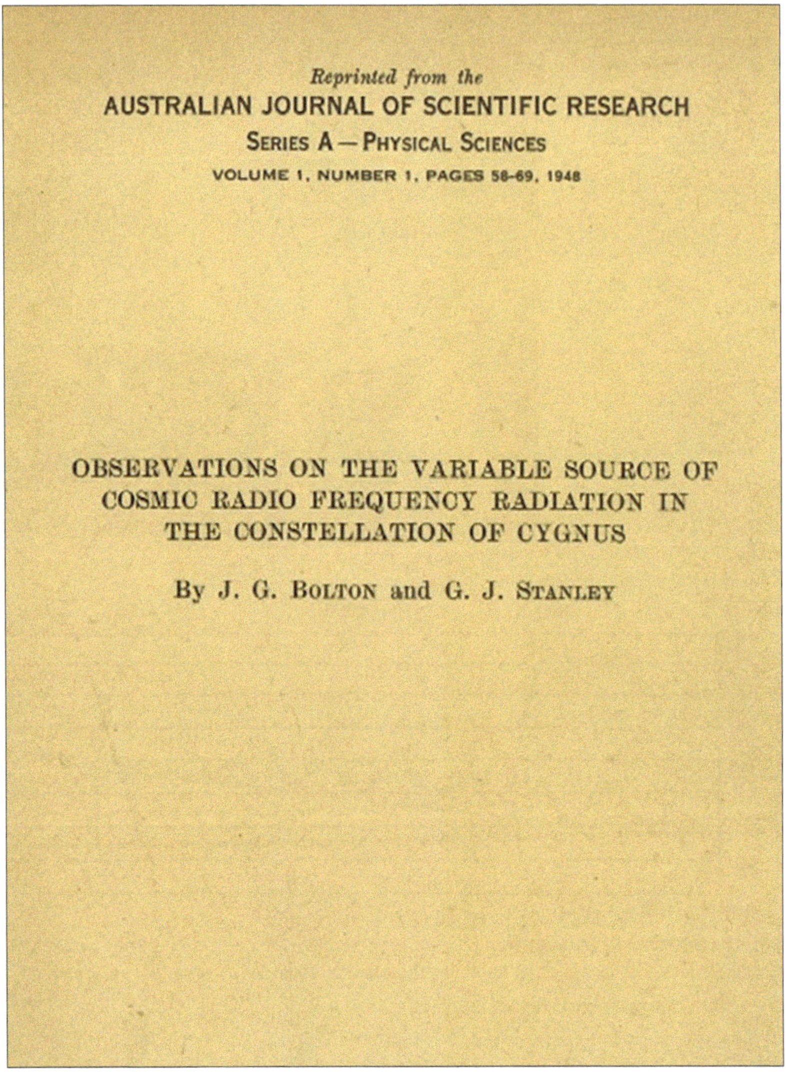

Reprinted from the
AUSTRALIAN JOURNAL OF SCIENTIFIC RESEARCH
SERIES A — PHYSICAL SCIENCES
VOLUME 1, NUMBER 1, PAGES 58–69, 1948

OBSERVATIONS ON THE VARIABLE SOURCE OF
COSMIC RADIO FREQUENCY RADIATION IN
THE CONSTELLATION OF CYGNUS

By J. G. BOLTON and G. J. STANLEY

Fig. 4.5 After their letter to *Nature*, Bolton and Stanley published a detailed account of the Cygnus work in the first issue of the new *Australian Journal of Scientific Research*. This provided the publication model at Radiophysics over the following years: rapid publication of a brief *Nature* paper to report a discovery and to establish priority, followed by a lengthy paper in the Australian journal to present and interpret the observational data (courtesy: Letty Bolton)

respectively 15 km and 35 km to the north (see Fig. 1.14), and measured intensity variations from the Cygnus source. They found there was a high correlation at the three sites, showing that the variations were either an intrinsic feature of the source itself or else were caused by an ionospheric or interplanetary diffraction pattern with electron density turbulence scale much larger than the 15 km spacing between the two nearest observing sites.

The pressure was now on to search for more radio stars, which led to a survey with the 100 MHz sea interferometer. Starting in November 1947, the observational work involved several nights on each of about 15 azimuth settings separated by about 10 degrees. If a weak interference pattern was suspected, the azimuthal position was checked several times before the existence of a source was accepted (the intensities of these new sources were about five times weaker than that of Cygnus). A second new radio star was found on 6 November in the Taurus constellation, and it was followed shortly after by two others in the Centaurus and Virgo constellations. See Orchiston and Slee (2006) and Robertson et al. (2010, 2014) for detailed accounts of the discovery of the Taurus and Centaurus sources respectively.

By the end of 1947 a fifth and sixth source had been discovered. To keep track of this growing list, Bolton decided to name each source after the constellation in which it was found, followed by a letter A, B, …to indicate that it was the strongest, second strongest source, etc. in that particular constellation. This naming convention was later adopted by radio astronomers around the world and it is still partly in use today.

These detections were quite some achievement given the primitive nature of the equipment, which

> … was very cranky … you got interference patterns one day and wouldn't get them the next. Equipment would fail … The sea interferometer had a lot of nasty habits, like you could get interference wiped out by refraction problems and get sources rising ten minutes of time late and all this sort of crazy stuff … (Stanley, 1974).

The positions obtained for these new sources were too inaccurate to attempt optical identifications, although Bolton, Slee and Stanley reasoned that they lay well beyond the Solar System and that they generated extremely-high levels of radio emission.

The Dover Heights group saw it as their first priority to obtain more accurate positions so that they could seek optical correlates for these radio stars, and what they required were much higher cliffs that would also allow observations of the sources when both rising and setting. They eventually selected two sites in the North Island of New Zealand, with coastal cliffs about 300 m above sea level, three times higher than the Dover Heights site. Stanley was responsible for designing the mobile laboratory on which four 100 MHz Yagis were mounted in sea-interferometer mode (Fig. 2.15). Between early June and early August in 1948, Bolton and Stanley carried out observations at Pakiri Hill and then Piha near Auckland (Fig. 4.6), often under appalling weather conditions, but the observations produced excellent interference fringes (e.g. see Fig. 4.7) which yielded much better positions for Cygnus A, Taurus A, Virgo A and Centaurus A. These were precise enough to allow tentative optical identifications for the last three sources with bright optical objects (see Table 4.1). Taurus A was associated with a Galactic supernova remnant known as the Crab Nebula, but the other two optical objects (NGC 4486 and NGC 5128 respectively) were peculiar (see Fig. 4.8), and as such it was not possible to

Fig. 4.6 The mobile sea interferometer at Pakiri Hill, New Zealand, in June 1948. The cabin mounted on the trailer could swivel horizontally to observe sources rising at different points along the horizon (courtesy: Stanley family)

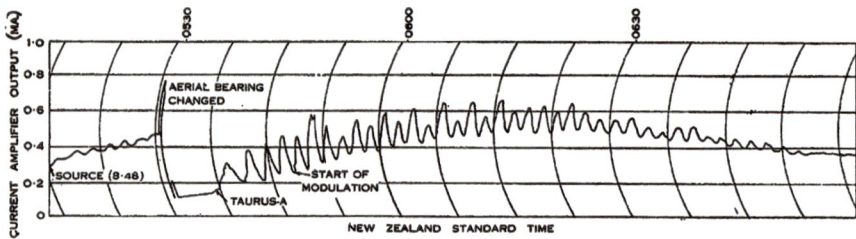

Fig. 4.7 The Taurus A interference pattern recorded at Pakiri Hill on 13 July 1948. Note the modulation of the interference pattern caused by another radio source in this region. Note also the absence of the spiky structure observed for Cygnus A in Fig. 4.4, a result of Taurus A being an extended source with angular dimensions of 4′ × 6′ (after Bolton and Stanley, 1949: 141)

Table 4.1 Three radio sources and their possible associated visible objects (adapted from Bolton, Stanley and Slee, 1949: 101)

Source	Position (Epoch 1948)		Possible associated visible object	
	R.A.	Dec.	Object	Remarks
Taurus A	05h 31m 00s ± 30s	+22° 01′ ± 07′	NGC 1952 (Messier 1)	Crab Nebula, expanding shell of an old supernova
Virgo A	12h 28m 06s ± 37s	+12° 41′ ± 10′	NGC 4486 (Messier 87)	Spherical nebula
Centaurus A	13h 22m 20s ± 60s	−42° 37′ ± 08′	NGC 5128	Nebula crossed by a marked obscuring band

Fig. 4.8 The first visible objects associated with radio sources by the Dover Heights group (from left): the Crab Nebula with Taurus A; NGC 5128 with Centaurus A; and M87 with Virgo A (courtesy: CRAIA)

determine whether they were Galactic or extragalactic in nature. To their disappointment, the Dover Heights group could not find any obvious optical object within the error box of the strongest radio star, Cygnus A. Despite this shortcoming, the RP Chief Taffy Bowen (1911–1991; 1973) suggested that the Bolton, Stanley and Slee (1949) paper in *Nature* was one of the most important ones published by RP in the early years of radio astronomy (e.g. see Fig. 4.9).

Actually, with the benefit of hindsight, we regard the *Nature* paper as *the* most important one published by RP in the early years of radio astronomy. It marked the birth of a new and dynamic branch of astronomy – extragalactic radio astronomy – which led to many of the exciting discoveries in astronomy during the second half of the twentieth century. For detailed studies of the discovery of the first discrete sources at RP see Robertson et al. (2014), Orchiston (2016c) and Robertson (2016, 2017).

One other important outcome of the New Zealand field trip was information about the enigmatic intensity fluctuations exhibited by Cygnus A (Fig. 4.4). Simultaneous observations of this source and the active Sun made by Slee at Dover Heights and by Bolton and Stanley in New Zealand showed conclusively that the Cygnus A variations were uncorrelated, while the solar bursts were in good agreement. Bolton, Slee and Stanley were now certain that the Cygnus A fluctuations were not intrinsic to the source, but were probably scintillations caused by diffraction in the intervening medium, with the scale size of electron density turbulence less than 2000 km. Unfortunately, Stanley and Slee (1950) did not publish this important result until late 1950. In the meantime, prompted by a letter from Pawsey (1949) to Martin Ryle, the Cambridge and Jodrell Bank groups carried out joint observations on Cygnus and, with a baseline of over 200 km, were easily able to establish an ionospheric origin for the fluctuations, with scale sizes from 5 to 10 km. The results were quickly published in *Nature* (Smith, 1950; Little and Lovell, 1950), but made no mention of the Radiophysics work (Sullivan, 2009: 324–27). When Lovell later learnt of the oversight he publically apologised to the Australians, but

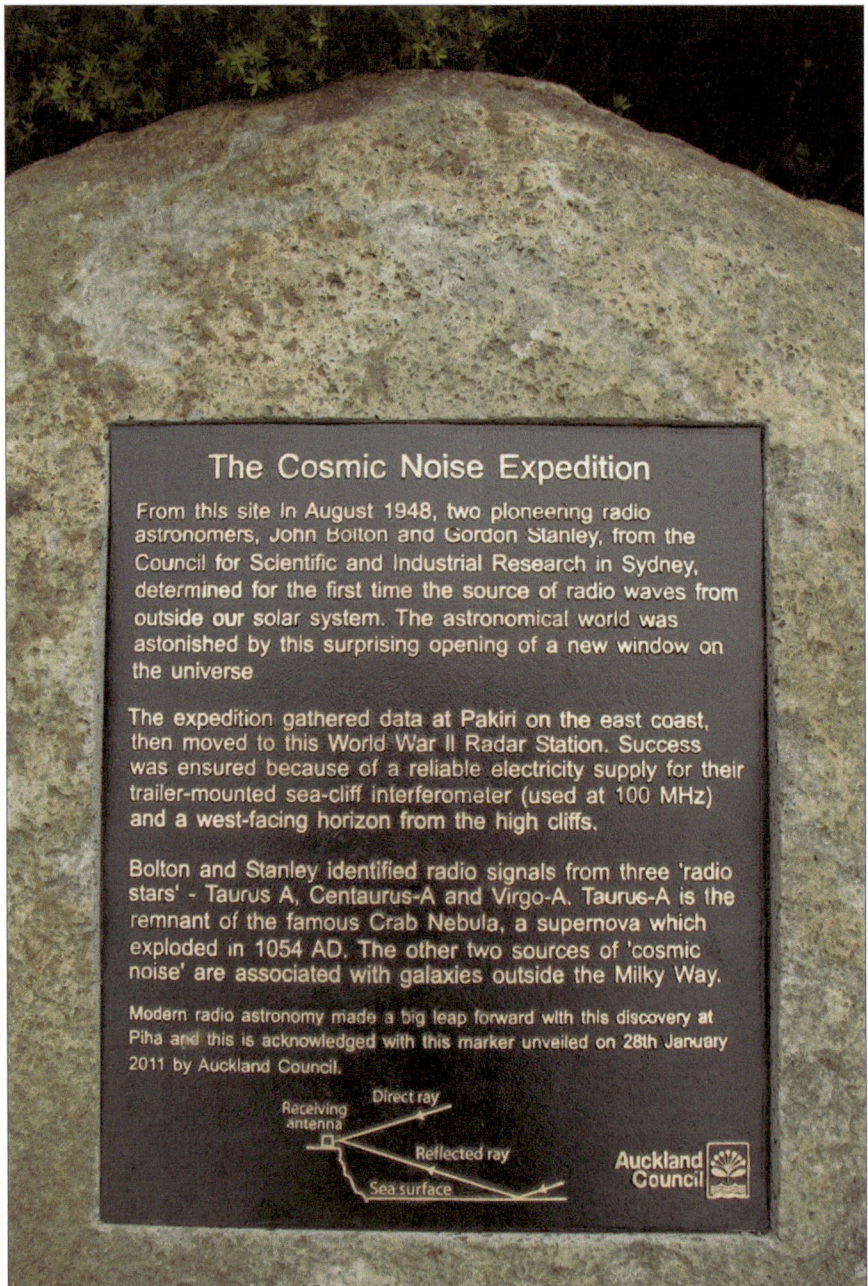

Fig. 4.9 Plaque to mark the visit by Bolton and Stanley to Piha on the west coast of New Zealand in 1948. A similar plaque at Pakiri Hill on the east coast was unveiled in February 2013 (courtesy: Miller Goss and Stanley family)

Fig. 4.10 The 9-Yagi array on its equatorial mounting was used to carry out the first survey of the southern sky at 100 MHz (after Pawsey and Bracewell, 1955: Plate 2)

the Cambridge group remained silent. This marked the beginning of an escalating tension between the Sydney and Cambridge groups (see below).

Following the New Zealand field trip, the search was on in earnest for more radio stars. During 1949 Bolton, Stanley, Slee and Kevin Westfold (1921–2001; Fig. 1.27) used the equatorially-mounted 9-Yagi array at Dover Heights (Fig. 4.10) to survey the sky at 100 MHz for new radio sources and to investigate the properties of known ones. This work produced 14 additional radio stars, including Hydra A, Hercules A, Fornax A and Pictor A. Stanley and Slee also used a number of 2-Yagi antennas to measure the intensities of the four strongest radio stars at a number of frequencies over the range 40–160 MHz, and published much more accurate source spectra (Fig. 4.11). These strongly suggested that the radio emission was non-thermal in origin. Bruce Slee (1978) later recalled that this was an exciting time to be involved in radio astronomy, for "… you could expect to find a new source almost every night." (see Bioboxes 4.1 and 4.2 for Slee and Stanley).

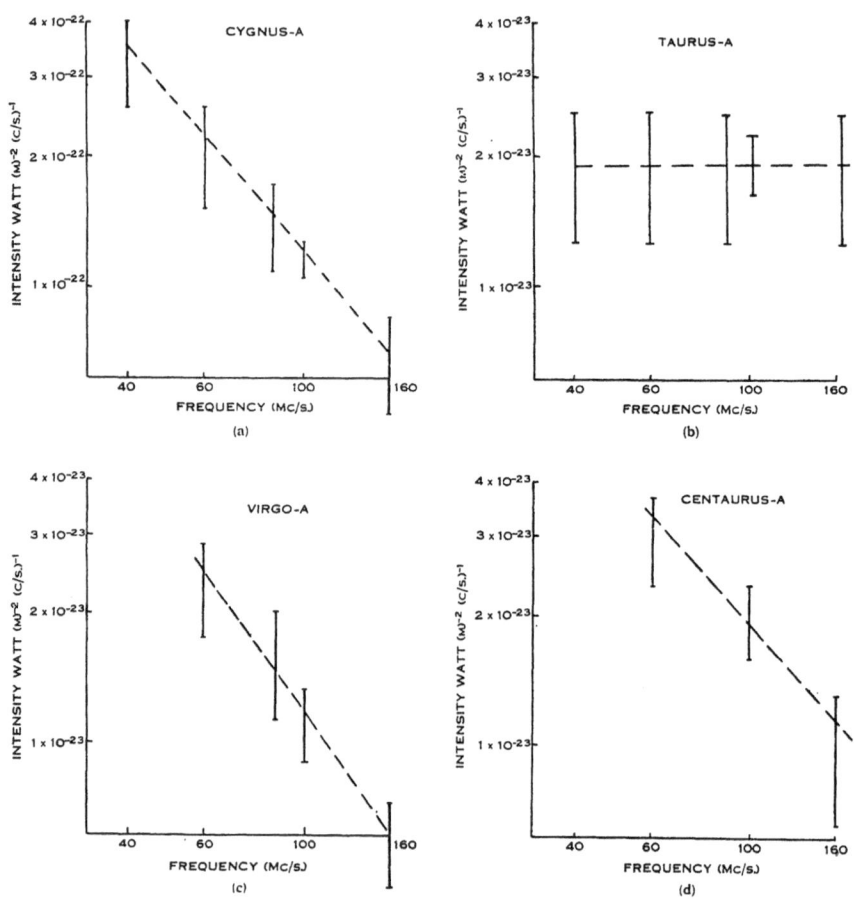

Fig. 4.11 The first radio spectra of the four strongest sources discovered at Dover Heights, over the frequency range 40 to 160 MHz. Cygnus A, Virgo A and Centaurus A show a sharp fall in intensity with increasing frequency. In contrast Taurus A shows a very flat spectrum, shown later to be a feature of radio emission from a small class of supernova remnants (after Stanley and Slee, 1950: 243)

Biobox 4.1: Bruce Slee
Owen Bruce Slee (Fig. 4.12) was born in Adelaide on 10 August 1924 (Orchiston, 2005; Orchiston, Sim and Robertson, 2016). He joined the Radiophysics Lab in 1946 after serving at radar stations during WWII and independently detecting solar radio emission at 200 MHz. Studying evenings, he completed a BSc (First Class Honours) at the University of New South Wales in 1959, and subsequently was awarded a DSc by the same university in 1971.

After serving with the Dover Heights team that elucidated the true nature of 'radio stars' (Slee, 1994), he gained further international recognition

(continued)

Biobox 4.1 (continued)

Fig. 4.12 Bruce Slee's career in radio astronomy spanned an extraordinary 70 years. His first paper was the pioneering letter to *Nature* reporting the detection and possible optical identifications of three new 'radio stars' (Bolton, Stanley and Slee, 1949) and his final paper was on the early development of the RP field stations (Orchiston and Slee, 2017) (courtesy: Bruce Slee)

during the 1950s through the Mills–Slee–Hill catalogue of radio sources at Fleurs, and the controversy this generated vis-à-vis the Cambridge 2C survey. Bruce then went on to pioneer the detection of radio emission from various types of active stars. Over the years he also made important contributions to pulsar astronomy and to the study of radio galaxies and clusters of galaxies. After retiring from the Australia Telescope National Facility in 1989 (as a Principal Research Scientist), he was made an ATNF Honorary Fellow and continued an active research program using the Australia Telescope Compact Array. He published extensively on a wide range of astronomical topics and, in recent years, also wrote a series of co-authored papers with Wayne Orchiston on the early history of Australian radio astronomy.

Bruce Slee "… is a pioneering Australian radio astronomer, and it is a testimony to his dedication and his passion for astronomy that after more than half a century he continues an active research program. Since joining the CSIRO's Division of Radiophysics in 1946 he has made an important, long-term, wide-ranging contribution to astronomy and to Australian science …" (Orchiston, 2004: 69). Bruce Slee died on the south coast of New South Wales on 18 August 2016. In 2016 the International Astronomical Union named minor planet 9391 after him, and in 2017 he was posthumously awarded a Member of the Order of Australia (AM) for his distinguished service to radio astronomy.

Biobox 4.2: Gordon Stanley

Gordon James Stanley (Fig. 4.13) was born in Cambridge, New Zealand, on 1 July 1921. After his family moved to Sydney, Gordon first worked for a manufacturer of electrical appliances and developed skills that would be essential to his future career. In 1943 he joined the Radiophysics Lab and developed expertise in designing and building receivers and antennas. In 1945 he received an engineering degree from Sydney Technical College and was promoted to Technical Officer at Radiophysics.

Early in 1947 Gordon joined John Bolton at the Dover Heights field station and they were able to show that intense radio emission from the Cygnus constellation came from a discrete, point-like source. Bruce Slee joined them late in 1947, and as a team they obtained accurate positions for three more of these discrete radio sources, Taurus A, Centaurus A and Virgo A, and this enabled them to identify the optical counterparts of these sources. This landmark achievement was published in a 1949 *Nature* paper (Bolton, Stanley and Slee 1949) and soon brought international recognition to the Dover Heights group and to Australian radio astronomy.

In 1955 Gordon and John Bolton were recruited by the California Institute of Technology (Caltech) in Pasadena to establish a radio astronomy group. They founded the Owens Valley Radio Observatory (OVRO) in northern California and Gordon designed the electronics for an interferometer consisting of two 27 m parabolic dishes (see Fig. 1.22). With Bolton's return to Australia in 1960, Stanley was subsequently appointed Director of the OVRO. During the 1960s he consolidated Caltech's position as one of the world's leading centres for radio astronomy.

After retiring from Caltech in 1975 he returned to his primary love of radio engineering. In 1982 he formed Stanley Microwave Systems which built microwave instruments for atmospheric and oceanographic applications. He spent his final years in Carmel Valley with wife Helen, where he died on 17 December 2001. For a memoir see Kellermann, Orchiston and Slee (2005).

Fig. 4.13 Gordon Stanley was an expert in the design and construction of receivers and antennas. The improvements he made to the receivers at Dover Heights led to the discovery of three of the first four discrete radio sources and their optical identifications (courtesy: Stanley family)

At the Potts Hill field station two other RP staff, Bernie Mills (1920–2011; Fig. 2.52) and Adin Thomas, were particularly interested in measuring accurate positions for the strongest radio stars, and they approached this challenge from another direction. Between May and December 1949 they carried out observations with the 97 MHz position interferometer (Fig. 2.39) developed earlier for solar work by Ross Treharne (1919–1982) and Alec Little (1925–1985) (Orchiston and Wendt, 2017). Mills and Thomas used the times of transit to establish the right ascensions of these sources and the periodicity of the interference patterns to determine their declinations. After applying various corrections they derived co-ordinates of R.A. 19h 57m 37 ± 6s and Dec. +40° 34 ± 3′ for Cygnus A, but were disappointed that within this error box "… there is no prominent or unusual celestial object with which the source can be identified" (Mills and Thomas, 1951: 170). However, they did note the presence of a faint extragalactic nebula just outside their error box (3′ away). They also found that the radio star was less than 3 arcmin in size, and that its characteristic fluctuations in intensity could be explained by terrestrial atmospheric effects. This was Mills' first excursion into non-solar astronomy, and followed lengthy discussions with Pawsey about fruitful new research prospects (see Sullivan, 2009: 337–339).

Mills continued this investigation when he moved to the Badgery's Creek field station, motivated by nagging concerns about the optical identifications suggested by Bolton, Stanley and Slee for Centaurus A and Virgo A: "… we always felt in the back of our minds that they may not be exactly right. This is one of the reasons I … built bigger antennas with the idea of getting more accurate positions with a larger number of sources" (Mills, 1976). Between February and December 1950 and for a short interval in 1951 he obtained further positional measurements for six prominent radio stars using three 101 MHz broadside arrays (Fig. 2.53). The results are listed in Table 4.2. The positions obtained for Taurus A, Centaurus A and Virgo A actually reinforced the optical identifications suggested by the Dover Heights group and confirmed the extragalactic nature of the last two sources, but it was the position obtained for Cygnus A that had greatest impact. Mills' value was published in the *Australian Journal of Scientific Research* (Mills, 1952a) and it coincided with the position published in *Nature* in 1951 by Graham Smith (b. 1923). Moreover, within the error boxes of both sets of values was a faint extragalactic nebula (the same one that Mills and Thomas had noted just outside their earlier error box), and closer examination of this object by Walter Baade (1893–1960) and Rudolph Minkowski (1895–1976) with the 200-inch Palomar Telescope in California (Fig. 4.14) revealed what appeared to be two faint colliding galaxies (see Fig. 4.15) (Baade and Minkowski, 1954).

Table 4.2 Positions of early radio stars, based on observations at Badgery's Creek (after Mills, 1952b: 461)

Source	Right Ascension (1950)	Declination (1950)
Centaurus A	13h 22m 30 ± 4s	−42° 46 ± 2′
Cygnus A	19h 57m 44 ± 2.5s	+40° 35 ± 1.5′
Fornax A	03h 19m 30 ± 6s	−37° 18 ± 3′
Hydra A	09h 15m 46 ± 4s	−11° 55 ± 8′
Taurus A	05h 31m 29 ± 2.5s	+22° 00 ± 3′
Virgo A	12h 28m 15.5 ± 2.5s	+12° 44 ± 6′

Fig. 4.14 Leading US astronomers, Walter Baade and Rudolph Minkowski, were strong supporters of the RP radio astronomy program. They used the 200-inch Palomar Telescope – then the largest reflecting telescope in the world – to search for the optical counterparts of the first radio sources (sketch by Russell Porter courtesy: the Archives, California Institute of Technology)

Fig. 4.15 The Cygnus A source (at centre) coincided with a very faint and peculiar nebulous object that appeared to be two galaxies in collision (courtesy: Orchiston collection)

Now only the Fornax A and Hydra A sources remained to be optically identified, although Mills suspected that NGC 1316 was connected with the former. Later he used the benefit of hindsight to comment on this:

> Fornax A is a double source and although the centroid is nicely on the galaxy, in fact the fine structure is more distributed on one side than the other so that the interferometer position is weighted towards the position of the fine structure. We now know what's happening, but at that time it was not clear. (Mills, 1976).

By 1952 the extragalactic nature of Cygnus A was thus established, and the Virgo A and Centaurus A sources were also confirmed to be extragalactic nebulae. Only Taurus A was associated with an object inside our Galaxy. These 'radio stars' seemed not at all to be stars, but mostly 'radio galaxies', distant objects that generated extraordinary amounts of radio energy. Stanley Hey in England had originally proposed that the Cygnus source was a strange radio-emitting star in our Galaxy, and most other radio researchers agreed for many years, but this idea was now put to rest. It would take another decade before radio telescopes evolved to the point where they could indeed detect radio emission from stars other than the Sun – and, ironically, one of the leaders in this new field of radio astronomy was none other than 'radio star' pioneer Bruce Slee.

This change in understanding saw a commensurate shift in terminology, and 'radio stars' became 'discrete sources', a generic term that had no specific Galactic or extragalactic connotation. The research focus shifted to establishing the celestial distribution of discrete sources at various frequencies, and in Australia this fitted perfectly with the strategy of maintaining small, largely autonomous groups of scientists at a number of different field stations. Several research teams directed their attention to surveys of these discrete sources.

For example, between June 1950 and June 1951, Alex Shain (1922–1960) and research assistant Charlie Higgins conducted an 18.3 MHz source survey of the sky mainly between –4° and –60° declination using a 30-element array at the Hornsby Valley field station. Altogether, 37 different discrete sources were detected, some of which exhibited intensity fluctuations. Only approximate positions were obtained for the weaker sources, but when all 37 were plotted (Fig. 4.16), they exhibited a broad distribution. For further details of this research see Orchiston et al. (2015).

At this time Bernie Mills was based at Badgery's Creek, and in 1950 he carried out a 101 MHz survey of the sky between declination +50° and –90° with a view to studying the Galactic distribution of discrete sources. He detected 77 sources (Fig. 4.17), and concluded that there were two types: (1) stronger sources, which were predominantly Galactic in nature because they were clustered close to the Galactic Plane, and (2) weaker sources, which were more randomly distributed over the sky. These latter sources might have been either extragalactic sources or weak Galactic sources relatively close to the Sun. This was a very different general picture to the one obtained by the Cambridge radio astronomers, as Mills (1976) was later to reflect:

> [The Cambridge radio astronomers] thought in terms of point sources uniformly scattered over the sky, radio stars in fact. And in the south we had measured all these large angular sizes and it was obviously much more complex than this … They had missed the strong sources near the plane [of the Galaxy] because they were using wide-spaced interferometers … This was the basic cause of our arguments at that time with Cambridge …

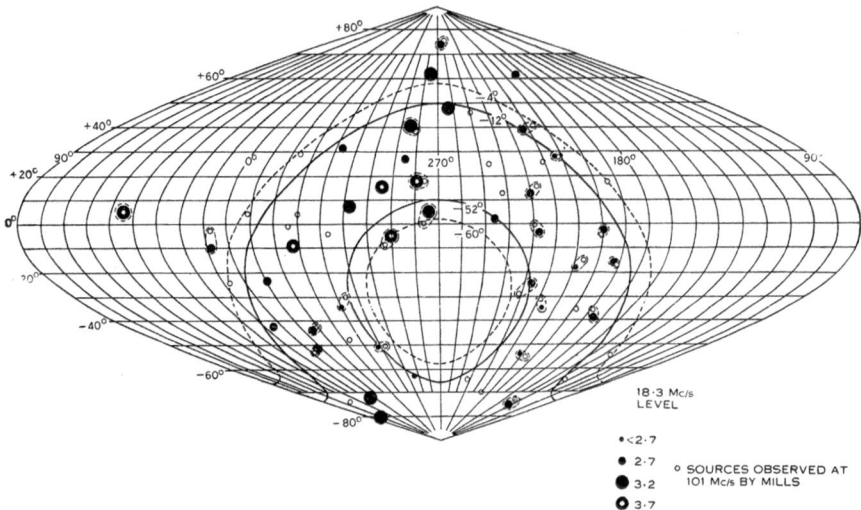

Fig. 4.16 The 18.3 MHz sources detected at Hornsby Valley in 1950–51 (after Shain and Higgins, 1954: 142). Sources marked by open circles are those that were also detected by Mills at 101 MHz (see Fig. 4.17).

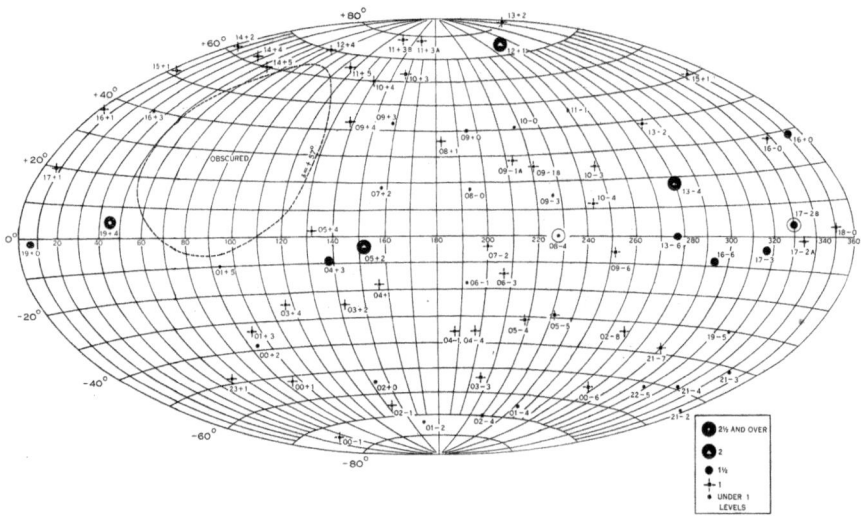

Fig. 4.17 The 101 MHz sources detected at Badgerys Creek in 1951 (after Mills, 1952a: 272)

At Dover Heights, John Bolton, Gordon Stanley and Bruce Slee carried out a comparable survey at the similar frequency of 100 MHz in 1951 using the new 12 Yagi array shown in Fig. 2.21 (Bolton et al., 1954a; see also Bolton et al., 1954b). According to Bolton (1978), this survey involved about two years' observing,

Table 4.3 Surveys of discrete radio sources 1950–54 (adapted from Bolton, Stanley and Slee, 1954a: 111)

Survey number	Observers (publication year)	Field station	Frequency (MHz)	Sensitivity limit (Jy)	Survey region (Dec.)	Number of sources
1	Stanley–Slee (1950)	Dover Heights	100	100	50° to –50°	22
2	Ryle–Smith–Elsmore (1950)	Cambridge	81	30	90° to 10°	50
3	Mills (1952a, 1952b, 1952c)	Potts Hill	100	50	50° to –90°	77
4	Hanbury Brown–Hazard (1953)	Jodrell Bank	158	5	70° to 40°	23
5	Shain–Higgins (1954)	Hornsby Valley	18	3000	10° to –90°	37
6	Bolton–Stanley–Slee (1954a)	Dover Heights	100	50	50° to –50°	104

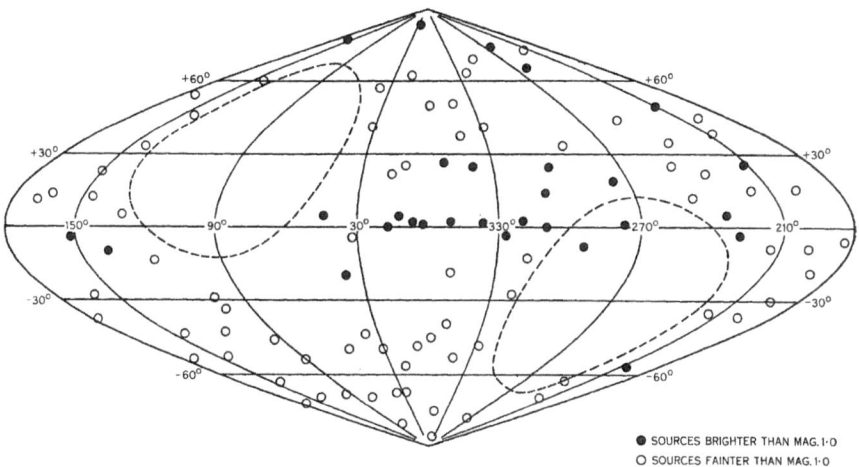

Fig. 4.18 The 100 MHz sources detected at Dover Heights in 1952–53 (after Bolton, Stanley and Slee, 1954a: 126)

largely because of problems generated by the ionosphere and thunderstorms. Sometimes "… there was only one night a week that you got good observations." Nonetheless, this work resulted in the detection of 104 sources, with flux densities down to 40 Jy. This was the most comprehensive list of radio sources available at the time, and showed that discrete sources were far from rare: increasing sensitivity seemed to bring increasing numbers of them (see Table 4.3 for a summary of the source surveys carried out over the period 1950–1954). The celestial distribution of the 104 sources is shown in Fig. 4.18, where the predominance of stronger sources along the plane of the Galaxy and of weaker sources in the southern polar cap is apparent. Meanwhile, new optical identifications suggested by Bolton, Stanley and

Slee only served to reinforce the view that most discrete sources were extragalactic in origin and associated with radio galaxies.

It is interesting that at this time there was little contact between the Dover Heights and Badgery's Creek groups. According to Mills (1976), this was because

> We each felt rather strongly our own technique was the best. Although we saw each other sometimes, Bolton lived out at Dover Heights and didn't come into the Lab very often and I spent most of my time out at Badgery's Creek. So we didn't have very much contact actually. And there were quite a few arguments about interpretation of the results and things like that.

The first attempt to look at discrete sources at much higher frequencies occurred in 1950 when Jack Piddington and Harry Minnett carried out a survey at 1210 and 3000 MHz from Potts Hill using the ex-Georges Heights radar antenna and two small parabolic antennas (see Figs 2.36 and 2.38) (Piddington and Minnett, 1951). They detected many of the stronger sources and, as a result of their Crab Nebula observations, Piddington later went on to theoretically investigate its associated magnetic field, coming up with a rapidly-rotating star. "That was the first, as far as I know, suggestion of a – what are they called now – rators?" (Piddington, 1978). What he was actually referring to was then known as a 'spinar'. Only two decades later would the neutron star (pulsar) at the centre of the Crab Nebula become known.

In 1953 Slee, Stanley and a new recruit to the Dover Heights group, Dick McGee (1921–2012), were able to learn something about discrete source emission at an intermediate frequency when they carried out a survey at 400 MHz using the 24.4 m 'hole-in-the-ground' antenna at Dover Heights (see Section 4.3 below). This small-scale survey concentrated on the Milky Way between declinations –17° and –49°, and the method of observation was described by McGee, Slee and Stanley (1955: 353):

> The Milky Way was observed on one fixed declination per day, changes in Right Ascension occurring as the rotation of the Earth swept the aerial beam across the sky. In a preliminary survey observations were made at intervals of 1° in declination. Later, in order to cover the more interesting regions in greater detail, intervals of ½° were used.

This survey produced a list of just 14 new sources, and with the benefit of hindsight and improved resolution we now know that about half of these were the result of confusion – instances where several adjacent sources were unresolved by the 2° beam of the antenna.

Further details of the 'hole-in-the-ground' research are presented in Orchiston and Slee (2002). For reviews of the various 'source surveys' conducted at Dover Heights and Potts Hill field stations see Orchiston and Robertson (2017) and Wendt et al. (2011b), respectively. For a detailed overview of the history of worldwide research on discrete sources up until 1953 see Sullivan (2009).

4.2   From Dover Heights to Fleurs

A major advance in the Australian study of discrete sources took place once the 85.5 MHz Mills Cross at Fleurs (Fig. 2.59) became operational in 1954. This new radio telescope was used for three different projects. Soon after its inauguration,

Bernie Mills, Alec Little and Kevin Sheridan (1918–2010; Fig. 3.23) searched for radio emission from specific types of celestial objects, including known novae, supernovae, and planetary nebulae. Unfortunately, they were spectacularly unsuccessful, the only object to reveal itself as a radio source being the Kepler supernova remnant of 1604. Nonetheless, Mills et al. (1956b: 84) boldly suggested that "… galactic radio emission near the plane of the Milky Way could be largely the integrated emission of supernova remnants …".

Mills and his colleagues had more success when they tried to observe 14 emission nebulae in our Galaxy, the Magellanic Clouds, and 11 bright southern galaxies with the new Cross. Six of the nebulae were detected in emission and one in absorption, and crude isophote maps were prepared for two of these (Fig. 4.19). Both Magellanic Clouds were detected (Fig. 4.20), as were eight of the 11 galaxies. Mills (1955: 369) found that these observations and others made with the Mills Cross in the general vicinity of the Galactic Centre

> … tend to support the ideas of [Russian theorist] Shklovsky [1916–1985] in which the galactic emission is considered to be distributed in two subsystems, one showing a discoidal distribution highly concentrated towards the galactic plane, and the other a very dispersed and approximately spherical distribution concentric with the galactic centre.

As we shall see, Mills would return to this 'disk–corona' dichotomy in later publications.

Sheridan (1958) also used the Mills Cross to study some of the strongest known radio sources, and was able to produce isophote maps for Centaurus A, Fornax A and Puppis A (Fig. 4.21). The Centaurus A and Fornax A results nicely complemented similar 19.7 MHz plots published at the same time by Shain (1958), based on observations made with the new Shain Cross at Fleurs (Fig. 2.63), and one of these is included here in Fig. 4.22 (left). For comparison Fig. 4.22 (centre) shows the 1400 MHz Centaurus A isophote plot that was produced at this time by Jim Hindman (1919–1999) and American visitor Campbell Wade (b. 1930) using the 11 m parabolic antenna at Potts Hill (Fig. 2.43).

One of the most unexpected findings at 85.5 MHz was that Centaurus A extends far beyond the boundaries of its associated optical galaxy. Furthermore, when Wade subtracted the intense localised central source he was left with two broad maxima, as shown in Fig. 4.22 (right). Upon reviewing the evidence at 1400 MHz, Hindman and Wade (1959: 268) came to a similar conclusion: "The source consists of two components: one is a point source evidently associated with the peculiar galaxy NGC 5128; the other is a very large extended source with no known optical counterpart."

The above-mentioned Mills Cross projects were interesting, but they were simply forerunners to the main work of the Cross which was to produce a detailed survey of the sky at 85.5 MHz. The observations for this were carried out by Bernie Mills, Eric Hill (1927–2016) and Bruce Slee between 1954 and 1957, and in the process they recorded about 2300 sources of discrete radio emission, publishing their results in a series of research papers in the *Australian Journal of Physics* (e.g. Mills and Slee, 1957). Although a number of the sources in this MSH Catalogue were associated with Galactic objects, the majority related to extragalactic nebulae. Edge and Mulkay (1976) and co-author Sullivan (1990) have described how this

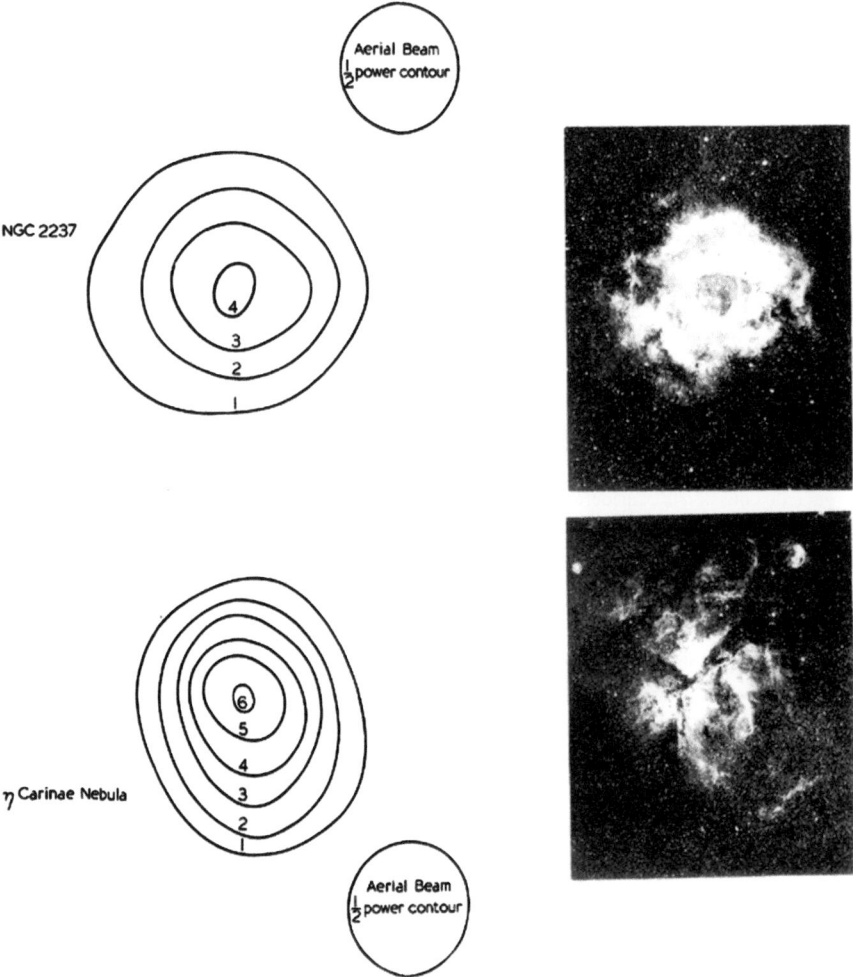

Fig. 4.19 Isophote plots at 85.5 MHz and associated optical images of the Rosette Nebulae (top) and Eta Carina (after Mills et al., 1956a: Plate 1)

had profound cosmological implications in terms of the competing 'Big Bang' and 'Steady State' theories which were prevalent at the time.

The MSH survey strongly disagreed with a parallel survey (called 2C) by Martin Ryle's (1918–1984; Fig. 1.39) group at Cambridge University. In the overlap region between the northern and southern surveys there was hardly any agreement between the very existence of individual radio sources (Fig. 4.23). Statistical studies of the histograms of source intensities also disagreed (Fig. 4.24).

According to Mills (1976), this issue had its origin when the prominent English astronomer Fred Hoyle (1915–2001) wrote a letter to the RP radio astronomers in

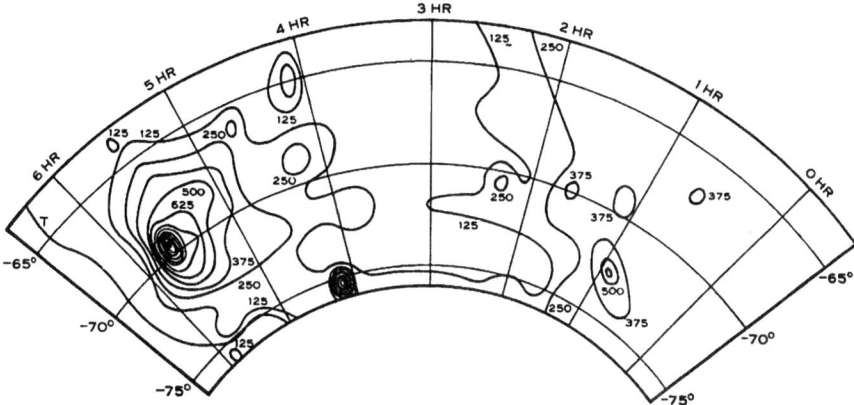

Fig. 4.20 Isophote plot at 85.5 MHz of the Large and Small Magellanic Clouds (after Mills, 1955: 372)

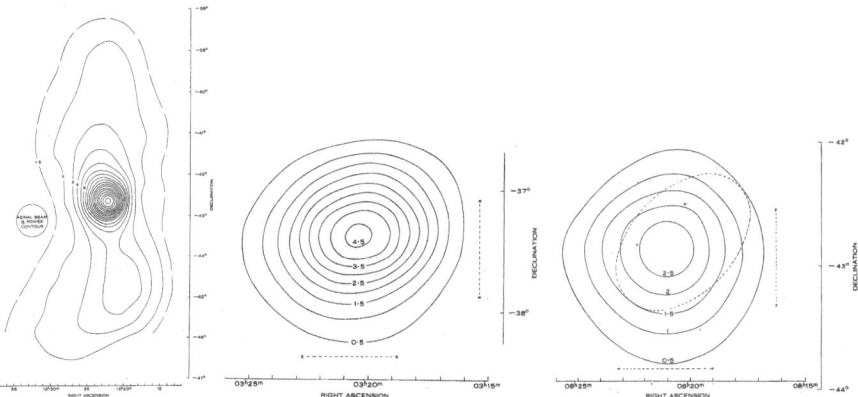

Fig. 4.21 Plots at 85.5 MHz of (from left): Centaurus A, Fornax A and Puppis A (after Sheridan, 1958: 404, 406 and 407)

which he mentioned "… a very great excess of faint sources which they had obtained at Cambridge, … and asking what did we get." At the time, Mills et al. had only preliminary results – which could not possibly agree, as he responded to Hoyle. When they examined the 2C evidence, Mills and Slee noticed that the Cambridge radio astronomers "… were not listing radio sources as such, because one could take an area where there were some quite strong sources in the Cambridge catalogue, look at our records, and see that there was absolutely nothing there."

Mills and Slee immediately came up with resolution effects as the explanation, but when Mills tried to raise this with Ryle his letters went unanswered. Later he was to remark: "We were a bit fed up with the Cambridge attitude at this time, I might say … They just ignored us. So we went ahead and did what we felt had to be

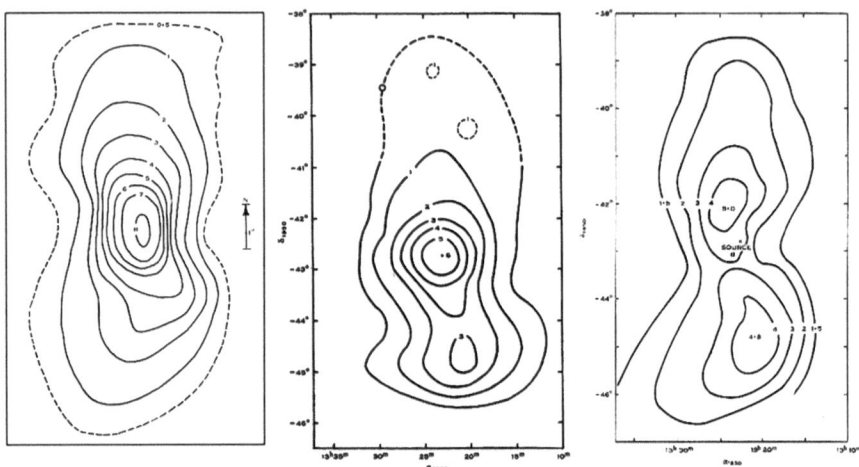

Fig. 4.22 Plots of Centaurus A (from left): at 19.7 MHz (after Shain, 1958: 520); at 85.5 MHz, following subtraction of the main central source (after Wade, 1959: 474); and at 1400 MHz (after Hindman and Wade, 1959: 266)

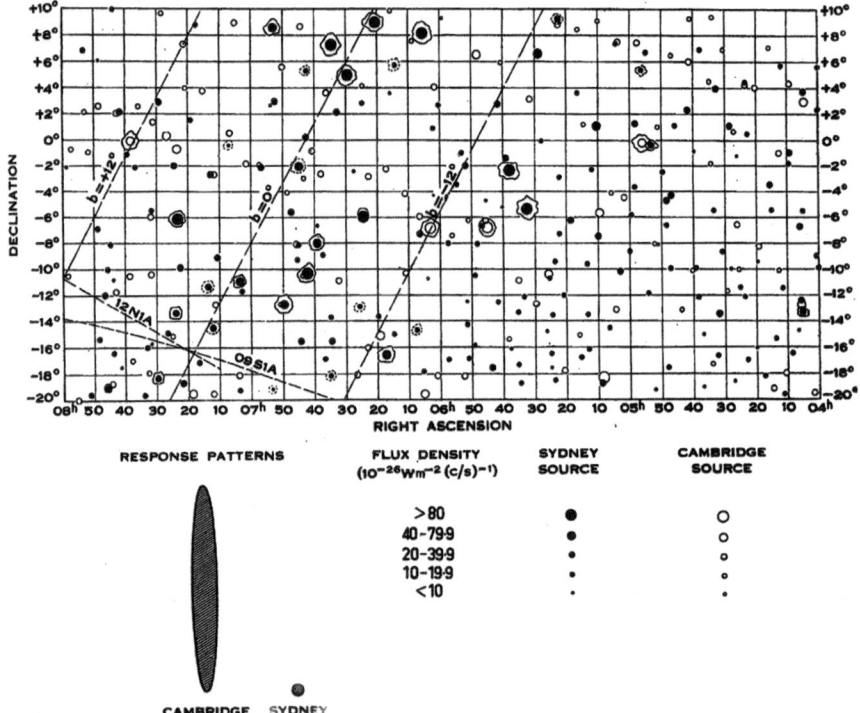

Fig. 4.23 Comparison of sources listed in the MSH Catalogue (closed circles) and the 2C Catalogue (open circles) for a selected area of sky. 'Extended' or 'large' sources in either catalogue are surrounded by an irregular line. The poor agreement between the two catalogues was at the heart of the Sydney–Cambridge controversy (after Mills and Slee, 1957: 168)

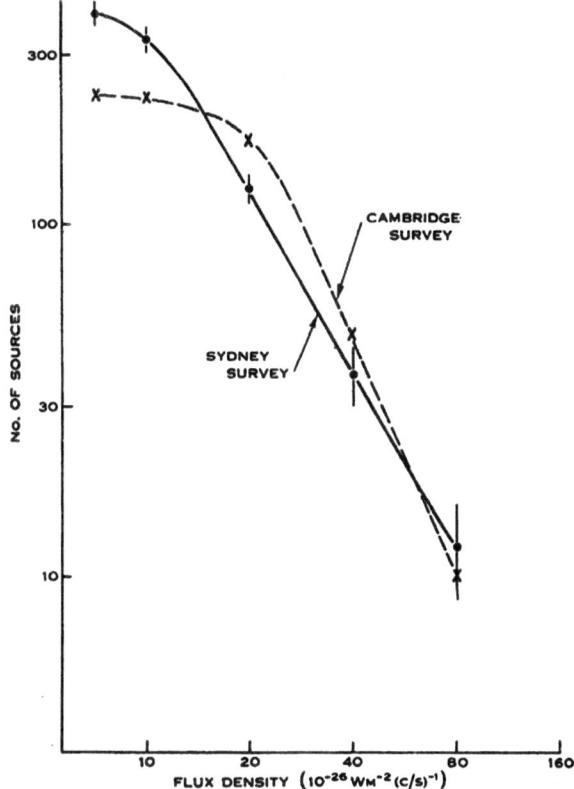

Fig. 4.24 The number counts at different source intensities for the Cambridge (2C) and Sydney (MSH) surveys produced discrepant curves. The steeper Cambridge curve was argued to favour the Big Bang over its rival Steady State cosmology (after Mills and Slee, 1957: 176)

done." (Mills, 1976). This was to publish a paper reporting the MSH–2C discrepancies and explain them largely in terms of shortcomings associated with the Cambridge interferometer.

A heated controversy erupted, and others who examined the evidence drew their own conclusions. By this time, John Bolton (1955a) had taken up a position at Caltech, and although he thought Ryle's work looked convincing on the surface he was quick to note that "… interferometers can be very misleading without some cross-check." Later in the year he had a chance to look over the Cambridge equipment and results, and "… I was horrified at their records and the amount of information that they had tried to dig out of the confusion." (Bolton, 1955b). Joe Pawsey and RP solar astronomer Paul Wild (1923–2008) were at a Jodrell Bank symposium in 1955 when Martin Ryle presented a paper on his source counts data, and Wild (1978) later remembered Pawsey

> … doing a magnificent rebuttal in a way, very modestly, on the basis of Mills' preliminary results. And Pawsey's performance reminded me of Marc Antony's speech, "Brutus is an

Fig. 4.25 Robert Hanbury Brown (right) with Taffy Bowen during the 1952 URSI congress. In 1962 Hanbury Brown was appointed to a chair in astrophysics at the University of Sydney where he spent the remainder of his career (courtesy: CRAIA)

honourable man", in reference to Martin Ryle. But he absolutely won the day, and I've never seen better form from Pawsey, who was normally such a quiet, unassuming, non-confrontational individual.

The Manchester radio astronomers also had problems with the Cambridge results. In January 1957, Robert Hanbury Brown (1916–2002; Fig. 4.25) – ever the diplomat – informed Pawsey that "Our present results suggest that there is a great deal wrong with the Cambridge survey, *but please do not quote me.*" (Hanbury Brown, 1957; our italics).

Yet not everyone saw things this way. David Edge (1932–2003) who joined the Cambridge group after the completion of the 2C survey, later retrospectively commented: "The fact is that *both* surveys were severely limited – one by confusion [the 2C], the other by sensitivity [Mills Cross] … the two surveys were simply incompatible …" (Edge, 1975). Mills later commented on the many extended sources in his survey (but not in Ryle's):

… we know now [1976] that a lot of these extended ones were [Galactic] background irregularities. We thought at the time they might be, but we included all these in the catalogue because, again, we didn't know what we had to look for at that time. And any concentration of emission, of any sort, appeared to be worth cataloguing. (Mills, 1976).

Despite these MSH shortcomings, in the end it took some years before the more serious problems associated with the Cambridge survey were fully recognised and the Australian results were largely accepted. In hindsight, it is now clear that the instrumentation used in *both* surveys came with its own inherent problems.

The Mills Cross survey also came at a time when astronomers were busy trying to explain discrete sources. Back in 1953, when Mills visited Minkowski and Baade at Caltech,

> It was all colliding galaxies … [and] there were a lot of discussions about probability of collisions, and how many there should be. There were some worries because there didn't seem to be enough collisions [to account] for the radio sources which were observed. There was very little physics in that part of it. It was purely looking for abnormalities in galaxies which might be identified with radio sources. But no one had a clue about the physical connection. (Mills, 1976).

Confusing the issue was the fact that normal galaxies – like our own and the Great Nebula in Andromeda – were also (weaker) sources of radio emission, so not all galaxies that generated radio waves were 'abnormal'. Furthermore, it was obvious to Mills and colleagues that their extended sources could be either Galactic or extragalactic, and might be due to blending of sources: "You see, at that time we knew about radio galaxies, we knew about clusters of galaxies, so we thought … it was highly probable that some of these extended sources did, in fact, represent blends of radio galaxies." (Mills, 1976).

And what about the prodigious energies involved? What were the associated emission mechanisms? By 1954 Mills had read papers by the Swedish physicist Hannes Alfvén (1908–1995), and the Swedish/Norwegian astrophysicist Nikolai Herlofson (1916–2001), and suspected that synchrotron radiation was involved, but it was only when he was introduced to the papers of the Soviet astrophysicist Iosif Shklovsky – now regarded as classics – that everything suddenly fell into place. Much later he admitted that in the mid-1950s "... my outlook was quite close to the Russian [one]. I was very impressed with the Shklovsky work … I think I was just generally thinking along similar lines." (Mills, 1976).

Concurrently with the MSH survey, RP's Alan Carter carried out an interferometric study of the sizes of selected Mills Cross sources using broadside arrays at Fleurs and Badgery's Creek (Fig. 2.53). Carter (1956) reported that

> From these observations we have about 50 reasonably reliable sizes, and about another 25 less reliable sizes. Many of these sources are about a minute or two of arc in diameter … but we do not think these sources are a representative sample of all sources down to our sensitivity level.

This inspired further studies, and in October 1956 Mills prepared a document titled "Proposed Source Size Measurements at Fleurs", in which he described the need for a radio-link interferometer involving part or all of the Mills Cross and a remote array at a more distant site, Wallacia, in order to measure the sizes of yet smaller (in angular size) MSH sources. He elaborated:

> The bread and butter aim of these measurements is to aid the identification of sources. Without angular size data it is often impossible to decide between several nebulae which may be within the limits of error in the position of a radio source. The provision of approximate angular sizes for the 1000 strongest sources could easily make the difference between 50 new identifications and 300. This difference arises because, without angular size data, one can only be reasonably certain of an identification between a bright source and a bright

nebula; with size data, one can hope for bright source–faint nebula and faint source–bright nebula identifications.

More interesting for a long term program is the possibility of recognising the very distant and powerful sources which are too remote for the associated galaxy to be detected optically … these must be very rare objects, but their recognition and the measurement of their properties can provide cosmological information which is unattainable by any other means …

Finally it is possible that the angular size interferometer proposed will itself be capable of making some significant advances in cosmology … (Mills, 1956a).

This project would provide for the continued use of the Mills Cross long after completion of the MSH source survey.

The first new investigation of source sizes was conducted by Bruce Goddard, Arthur Watkinson and Bernie Mills in 1958 and 1959, and used the E–W arm of the Mills Cross, a 91.4 m section of the S arm of the Cross, and an identical 50-dipole N–S array at Wallacia (Fig. 4.26) (Goddard et al., 1960). Drawing on a sample of about 1200 sources, mostly from the MSH survey, they were able to use source size in conjunction with other data to suggest identifications for about 10% of the sources. Mills elaborated:

A total of 46 possible identifications with galaxies are listed and 55 possible identifications with clusters of galaxies, the great majority of which are new. Most of these galaxies are double systems … [but] it seems probable that many galaxies of completely normal appearance are very strong radio emitters … In many cases, the emission from clusters appears to be associated with a single galaxy or pair of galaxies in the cluster … (Mills, 1960: 550).

Mills then moved to the University of Sydney and with more pressing demands on his time it was left to Bruce Slee and visiting Cambridge radio astronomer, Peter Scheuer (ca.1930–2001), to take the source size project further. In 1961 and 1962 they used the E–W arm of the Mills Cross in conjunction with 'barley-sugar' arrays (see Fig. 4.27) erected at four different remote sites to the north and south of Fleurs to investigate a large number of MSH sources (Scheuer et al., 1963). Scheuer was responsible for the final reduction of the observations and he took all the chart records with him when he returned to England, but unfortunately never found the

Fig. 4.26 The 50-dipole N–S array under construction at the Wallacia field station (courtesy: CRAIA)

Fig. 4.27 Peter Scheuer adjusting one of the 'barley-sugar' antennas at the Freeman's Reach remote site in 1962 (courtesy: CRAIA)

time to properly analyse the observations. Consequently, few results from this investigation were published, and Scheuer later admitted that this was a source of considerable embarrassment to him.

While Goddard and his collaborators were engaged in the above-mentioned source size project, Richard Twiss (1920–2005), Alan Carter and Alec Little (see Biobox 4.3) used various Chris Cross aerials as a 1427 MHz interferometer in order to investigate the structure and polarisation of a number of bright southern radio sources. Among other findings, they were able to show that Centaurus A consists of two distinct components, "… each of about 2½′ size between half-intensity points and separated by 5′." (Twiss et al., 1960: 156), and that Taurus A was roughly elliptical in shape.

Biobox 4.3: Alec Little

Alec George Little (Fig. 4.28) was born in Sydney on 2 February 1925, and left school at age 15 to join the Radiophysics Lab. He soon became a junior laboratory assistant and then made rapid progress through the technical staff ranks. Alec was "… a keen and gifted technician … and was generally regarded as a young man who was going places" (Mills, 1985: 113). After WWII he worked briefly in the valve laboratory before transferring to the radio astronomy group, where his career would flourish.

In collaboration with Ross Treharne and then with Ruby Payne-Scott, Little developed an innovative solar interferometer (Wendt and Orchiston, 2019), and he then worked with Bernie Mills on the Mills Crosses at Potts Hill and Fleurs. Studying part-time at Sydney Technical College, he completed Certificate and Diploma courses before obtaining a BSc (First Class Honours) from the University of New South Wales in 1955. He then spent two years at Stanford University as a Research Associate, in the process obtaining an MSc

(continued)

Biobox 4.3 (continued)

Fig. 4.28 Alec Little left school at 15 and joined Radiophysics as a messenger boy. He studied part-time and steadily worked his way up into the research ranks (courtesy: CRAIA)

in radio astronomy. When Bernie Mills joined the University of Sydney in 1960, Alec followed him there, accepting a lectureship in the School of Physics. He and Mills worked on the Molonglo Cross (Fig. 5.5), and in 1968 Alec was appointed Director of the Molonglo Radio Observatory and in 1985 an Associate Professor. Subsequently he was responsible for overseeing the conversion of the Cross into the Molonglo Observatory Synthesis Telescope (see next chapter). During this period, Alec was also associated with the Fleurs Synthesis Telescope.

In the early 1980s he was appointed a consultant to the CSIRO Australia Telescope project. A former President and Vice-President of the Astronomical Society of Australia, he also served as the Australian representative on URSI Commission 5 (Radio Astronomy). Alec Little played a major role in the development of Australian radio astronomy and "His cheerful enthusiasm, forthright personality and lack of any pretensions endeared him to all his colleagues …" (ibid.). He died suddenly from a heart attack on 20 March 1985.

With the installation of a 60-ft dish at the eastern end of the Chris Cross at Fleurs, the RP radio astronomers had access to a powerful new instrument, the Fleurs Compound Interferometer (Fig. 2.66), with which to study southern radio sources at 1440 MHz, and two teams took up this challenge. Norman Labrum (1921–2011), T. Krishnan (b. 1933), Warren Payten and Eric Harting observed eight well-known Galactic and extragalactic sources, deriving angular sizes and accurate positions for

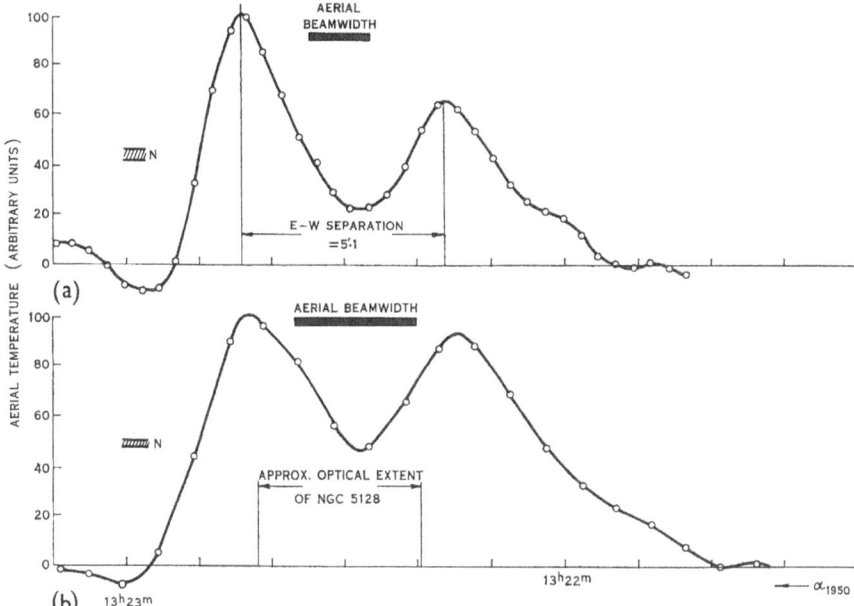

Fig. 4.29 Centaurus A profiles obtained with the Fleurs Compound Interferometer with beam widths of (a) 1.58′ and (b) 3.1′, showing the double nature of the source (after Labrum et al., 1964: 329)

each. Further to the result reported by the Twiss team, they found the eastern and western components of Centaurus A to have sizes of 2.7′ and 2.1′, separated by 5.1′ (Fig. 4.29).

For their part, Don Mathewson (b. 1929), John Healey and John Rome plotted emission along the Galactic Plane, noting the presence of a chain of conspicuous discrete sources (Mathewson et al., 1962a). Some of these were thermal sources and associated with gaseous nebulae previously identified by optical astronomers, and others were non-thermal sources linked to supernova remnants. Orchiston and Mathewson (2009) provide further details of research carried out at Fleurs with the 60-ft dish, and a summary of its research after its transfer to Parkes is contained in Orchiston (2012).

4.3   Radio Emission from the Galactic Centre

Between May and November 1949, Alex Shain (Fig. 2.28) carried out a survey of radio emission from a zone of sky centred on −34° declination using the highly-directional 9.15 MHz transit array at the Hornsby Valley field station (Fig. 2.30).

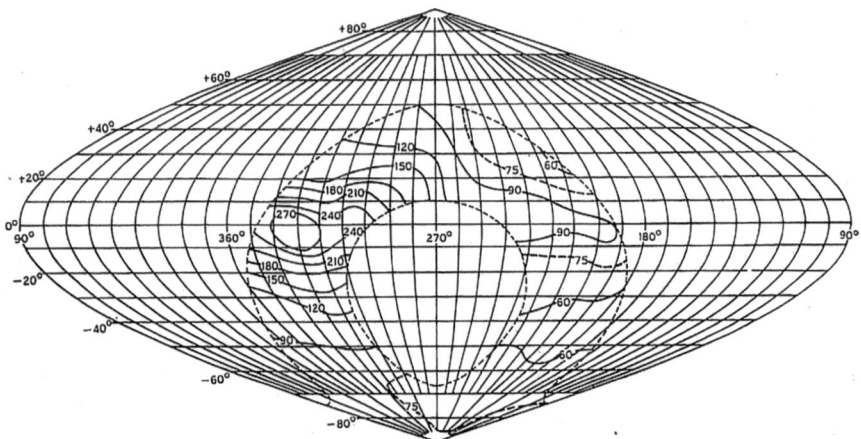

Fig. 4.30 Map of 18.3 MHz radiation in the general vicinity of the Galactic Centre (after Shain and Higgins, 1954: 137)

This zone of sky included the Galactic Plane and the Galactic Centre, and the resulting crude isophote plot showed enhanced radiation along the Plane with the highest flux levels in the vicinity of the Galactic Centre. Shain and Charlie Higgins (1954) were able to improve on this rather unexciting diagram between June 1950 and June 1951 when they carried out a more detailed survey with an enlarged 18.3 MHz array. Their new isophote plot (Fig. 4.30) showed a strong source in the region of the Galactic Centre. However, in reporting this interesting result they cautioned that the true intensity of this source may be under-represented because of gas absorption at 18.3 MHz in the direction of the Galactic Centre. For further details see Orchiston et al. (2015).

By the time Shain and Higgins belatedly published their results in 1954, several colleagues had already drawn attention to the distinctive discrete source located at the centre of our Galaxy. The first to do so were Jack Piddington and Harry Minnett who conducted preliminary observations at 1210 MHz with much better angular resolution, using a 3 m paraboloid at Potts Hill in 1948 and later the much larger ex-Georges Heights radar antenna at both 1210 and 3000 MHz in 1950. In their report, published in the *Australian Journal of Scientific Research* in 1951, they remarked on a "… new, and remarkably powerful, discrete source" (Fig. 4.31) at R.A. 17h 44m and Dec. −30° close to the assumed centre of our Galaxy. Piddington and Minnett (1951: 467) referred to this as "The Discrete Source in Sagittarius–Scorpius", given that its location was on the border of these southern constellations. By drawing on unpublished data provided by Shain (18.3 MHz) and Mills (100 MHz) they were able to plot its spectrum, which indicated "… an optically thin, thermally radiating gas …" (Piddington and Minnett, 1951: 469).

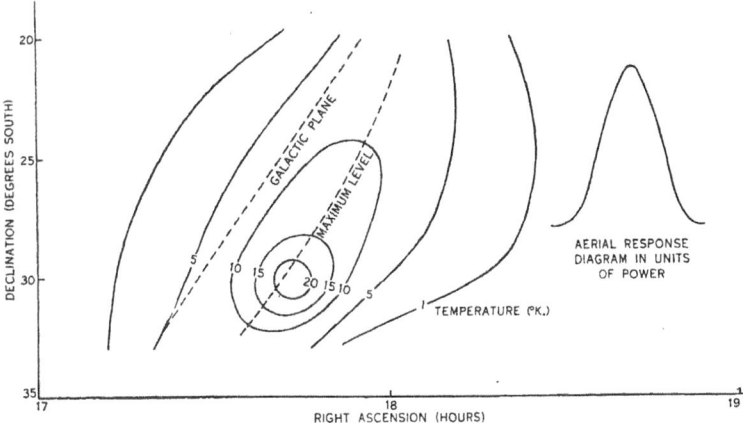

Fig. 4.31   1210 MHz crude contours in the vicinity of the Galactic Centre, indicating the presence of a discrete source (after Piddington and Minnett, 1951: 465)

Biobox 4.4: Dick McGee

Richard Xavier McGee (Fig. 4.32) was born in Sydney on 31 December 1921 and died on 19 December 2012 (Sim, 2013). During WWII he served in the army and in the air force. He later studied at the University of Sydney as a student with the post-war Commonwealth Reconstruction Training Scheme, obtaining a BSc with First Class Honours in 1950. He then joined Radiophysics and, after working at the Dover Heights and Murraybank field stations, he

Fig. 4.32   Dick McGee working on the primary feed of the 6.4-m Murraybank antenna (courtesy: CRAIA)

(continued)

Biobox 4.4 (continued)

became a stalwart of the group researching emission from hydrogen and interstellar molecules with the Parkes Radio Telescope.

In 1969 Dick was awarded a DSc by the University of Sydney in recognition of his published work. He was also interested in astronomy outreach and the history of Australian radio astronomy, and for a time served as RP's honorary historian. Along with Rosslyn Haynes, David Malin, and RP colleague Raymond Haynes, he produced the classic text *Explorers of the Southern Sky. A History of Australian Astronomy* (Cambridge University Press, 1996). A founding member of the Astronomical Society of Australia, Dick served as Editor of the Society's research journal from 1971 to 1988.

On Dick's retirement in 1986 his colleague and former Chairman of CSIRO, Paul Wild, wrote: "I want to thank you for your enormous contribution to the Division. Not only for your research contributions – the discovery of the centre of the Galaxy, pioneering work on HI, and huge works in unraveling the mysteries of galactic astronomy – but also for many other contributions which helped the world, and Radiophysics, to go round and be more enjoyable places …" (Wild, 1986).

At Dover Heights, John Bolton, Kevin Westfold, Gordon Stanley and Bruce Slee independently detected this Galactic Centre source in late 1951 when they carried out observations at 160 MHz with the 21.9 m prototype 'hole-in-the-ground' antenna (Fig. 2.22). Subsequent observations in 1953 by Slee, Stanley and new team member, Dick McGee (Biobox 4.4), with the concreted 24.4 m antenna at 400 MHz (Fig. 2.24) produced an impressive contour plot and it was they who coined the name 'Sagittarius A' for this source in papers published in *Nature* (see Fig. 4.33) and in the *Australian Journal of Physics* (McGee et al., 1955). Before these papers were published, Joe Pawsey sent a copy of the isophote plot to Walter Baade at the Mt Wilson and Palomar Observatories for his evaluation. His reply is illuminating:

> Frankly, I jumped out of my chair the moment I saw what it meant. I have not the slightest doubt that you finally got the nucleus of our galaxy!! … Altogether I concluded about two years ago – after a careful examination of my 48 inch Schmidt plates of the nuclear region of our galaxy and thorough checking of all suspicious objects at the 200 inch – that there was positively no chance whatsoever to detect the nucleus of our galaxy in the optical range and that we had to await what you radio people could do about it … It is very improbable that the coincidence between inferred and observed position of the nucleus is accidental. (Baade, 1954).

As it turned out, the position of Sagittarius A *was* particularly meaningful, and eventually was adopted by the International Astronomical Union to define the Galactic Centre and the anchor point for a new Galactic co-ordinate system. Sagittarius A is now widely known by its abbreviated name Sgr-A.

Further information on the discrete source at the centre of our Galaxy was forthcoming in 1955, when Bernie Mills, Bruce Slee and Eric Hill used the newly-constructed Mills Cross at Fleurs to survey the Galactic Plane at 85.5 MHz.

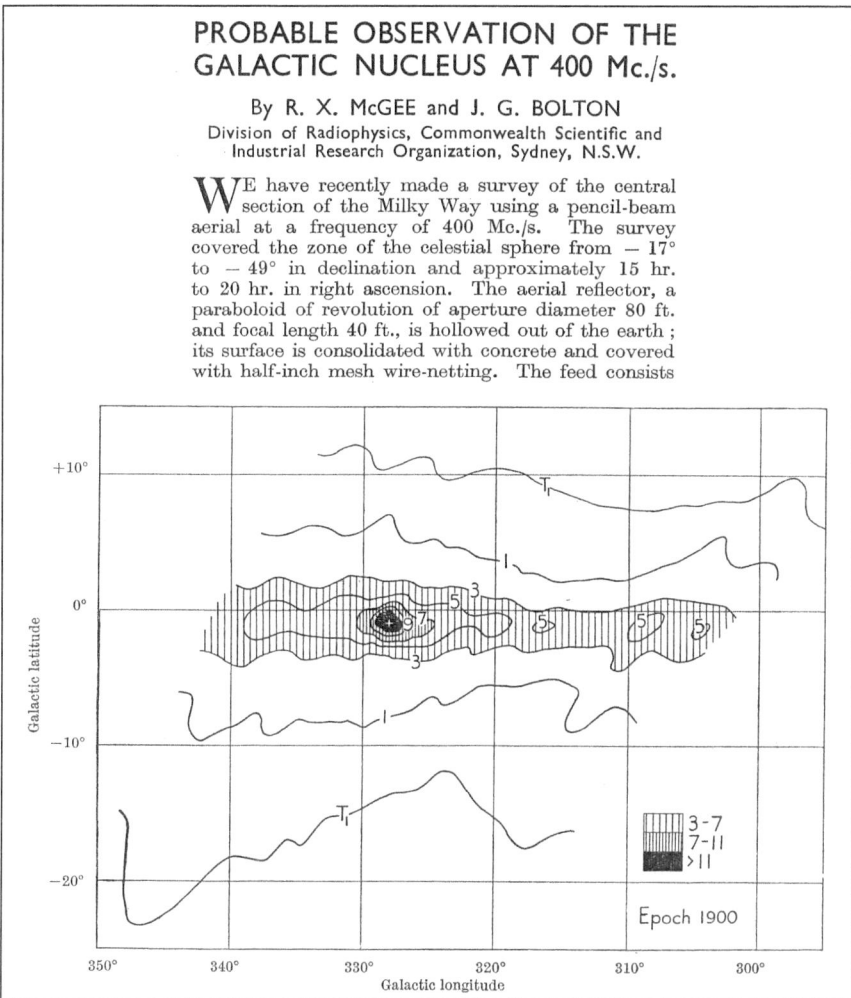

PROBABLE OBSERVATION OF THE GALACTIC NUCLEUS AT 400 Mc./s.

By R. X. McGEE and J. G. BOLTON

Division of Radiophysics, Commonwealth Scientific and Industrial Research Organization, Sydney, N.S.W.

WE have recently made a survey of the central section of the Milky Way using a pencil-beam aerial at a frequency of 400 Mc./s. The survey covered the zone of the celestial sphere from $-17°$ to $-49°$ in declination and approximately 15 hr. to 20 hr. in right ascension. The aerial reflector, a paraboloid of revolution of aperture diameter 80 ft. and focal length 40 ft., is hollowed out of the earth; its surface is consolidated with concrete and covered with half-inch mesh wire-netting. The feed consists

Fig. 4.33 A three-page paper by McGee and Bolton on the Galactic nucleus was published in *Nature* in May 1954. After rigorous internal refereeing at Radiophysics, the word 'probable' was added to the title to cover the possibility of a chance coincidence (after McGee and Bolton, 1954: 985)

The much smaller beam of this powerful radio telescope resolved Sagittarius A into two distinct components separated by about 2°, superimposed on a bright background that extended along the Galactic Plane for several degrees. In a paper published in *Observatory*, Mills (1956b) concluded that this may be due to absorption caused by an HII region situated in front of, or partially in front of, the Sagittarius A source. Mills (1976) later reflected:

This was obviously an intensely interesting result. It was one of the few occasions I think where I sat down and wrote a paper very rapidly as soon as we had the contours worked out. It was clear that this non-thermal emission was arising on either side of the thermal emission.

Evidence of absorption along the Galactic Plane at even lower frequencies emerged at about this time following the commissioning of the new 19.7 MHz Shain Cross at Fleurs. Initially, the main research program involved the observation of a strip of sky extending some 10° on either side of the Galactic equator, and of particular interest was the appearance of the Galactic Centre. A map of the region (Fig. 4.34) revealed that Sagittarius A was indeed seen in absorption.

In a useful review paper published in 1959, Mills brought together all the Australian observations of Sagittarius and concluded that although several physical models explain the derived data,

> … the most plausible one, and probably the correct one, consists of an HII region embedded in a flattened spheroidal non-thermal source, both being located at the galactic nucleus. The dimensions of the HII region would then be of the order of 89 pc × 35 pc and its mass some 2.5 × 10[5] times that of the Sun. (Mills, 1959b: 281).

The final study of the Galactic Centre region conducted by RP radio astronomers prior to their assault on this region with the Parkes Radio Telescope came in

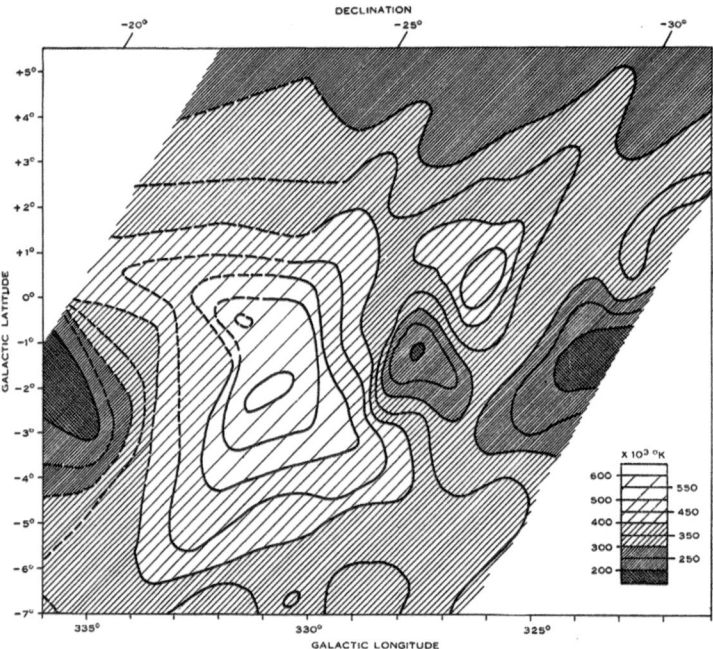

Fig. 4.34 Map at 19.7 MHz in the vicinity of the Galactic Centre showing clear evidence of absorption (after Shain, 1957: 197)

April–May 1961 when Mathewson, Healey and Rome used the newly-constructed Fleurs Compound Interferometer with its 50' beam. Their observations not only confirmed the discovery by the Dutch–American radio astronomer Gart Westerhout (1927–2012) of a ring of ionised hydrogen in the inner reaches of the Galaxy, but also suggested that near the Galactic Centre this hydrogen was "… concentrated in an irregular spiral structure … [with] peaks in the thermal emission at $l^{II} = 330°$, 338°, 14°, 27°, where the line-of-sight is tangential to the arms." (Mathewson et al., 1962b: 374).

For more on the discovery of the Galactic Centre see Morton (1985) and Robertson and Bland-Hawthorn (2014). It is interesting to note that Andrea Ghez (University of California) and Reinhard Genzel (Max Planck Institute) were awarded the Nobel Prize for Physics in 2020 for their discovery of a supermassive black hole at the Galactic Centre.

4.4   Surveys of the Background Radio Emission

The first map of background radio emission from the sky – as visible from Sydney – was published by John Bolton and Kevin Westfold in the *Australian Journal of Scientific Research* in 1950 (see Fig. 4.35). The map was the result of a survey carried out at Dover Heights with the 9-Yagi 100 MHz antenna on its equatorial mount (Fig. 4.10). Bolton and Westfold (1950: 19) pointed out that their survey and those conducted by Hey et al. at 64 MHz and Reber at 160 MHz collectively provided "… a complete picture of the noise distribution over the whole celestial sphere … [which] will undoubtedly be of value in studying correlation between

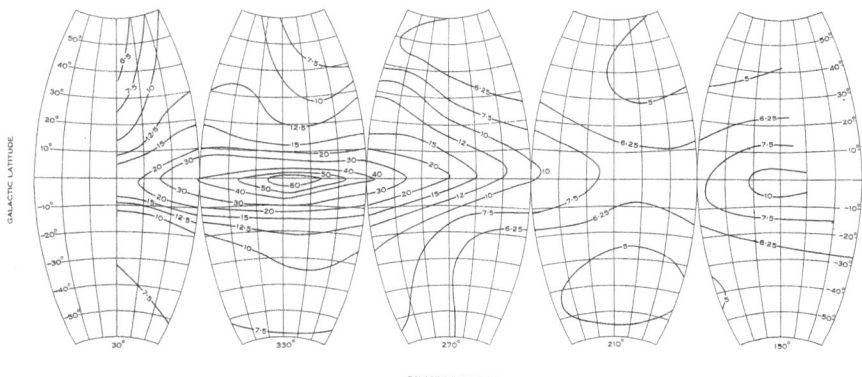

Fig. 4.35 The first Australian 'all-sky' map of 100 MHz radiation, obtained with the Dover Heights 9-Yagi antenna (after Bolton and Westfold, 1950: 39)

radio-frequency and optical data and possibly in deducing the form of the Galaxy."
The Sydney 100 MHz map showed that the strongest emission derives from the
region of the Galactic Plane, mainly between Galactic longitudes 270° and 20°.

Although Bernie Mills and his Fleurs colleagues were primarily interested in
discrete sources, their 85.5 MHz Mills Cross survey of the mid-1950s did provide
data on overall Galactic background emission. Mills (1959a: 431) was able to identify two discrete types of emission:

> (1) an extensive component of small axial ratio, displaying only moderate concentration
> toward the plane of the Milky Way and the galactic centre, which is here called the corona;
> [and] (2) a much brighter component, closely confined to the Galactic Plane and concen-
> trated in the inner regions of the Galaxy (the disk) ...

Fig. 4.36 shows the general distribution of the coronal emission, with the hatched
region indicating the Galactic Plane component. Mills was not able to ascertain
whether the non-uniformities in the coronal emission were primarily Galactic or
extragalactic in origin.

The first comprehensive survey made by RP scientists at a much higher frequency was carried out by Jack Piddington and Gil Trent at 600 MHz using the
11 m dish at Potts Hill. This radio telescope, which had a 3.3° beam, was used in
1954 to survey Galactic emission between declinations –90° and +51°. This was
one of the most comprehensive surveys conducted since Reber's pioneering efforts,
and Piddington was particularly proud of it. Piddington and Trent found – as with
previous 'all-sky' surveys – that emission was concentrated along the Galactic
Plane, where a succession of discrete sources is apparent (see Fig. 4.37). Piddington
and Trent (1956a: 82) thought it likely that most of these discrete sources "... are
thermally emitting clouds of ionised hydrogen."

It was only with the advent of the 19.7 MHz Shain Cross at Fleurs that RP radio
astronomers were finally able to gain an insight into the overall distribution of

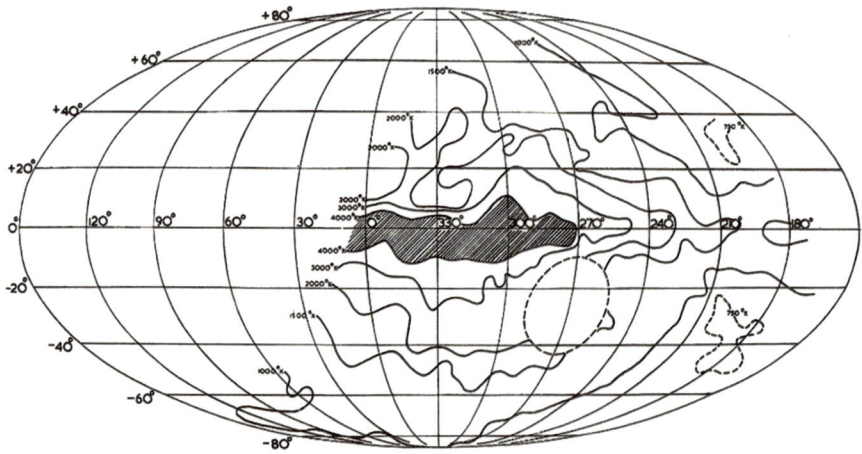

Fig. 4.36 Map of 85.5 MHz emission from the Mills Cross survey at Fleurs. The hatching indicates the Galactic Plane (after Mills, 1959a: 433)

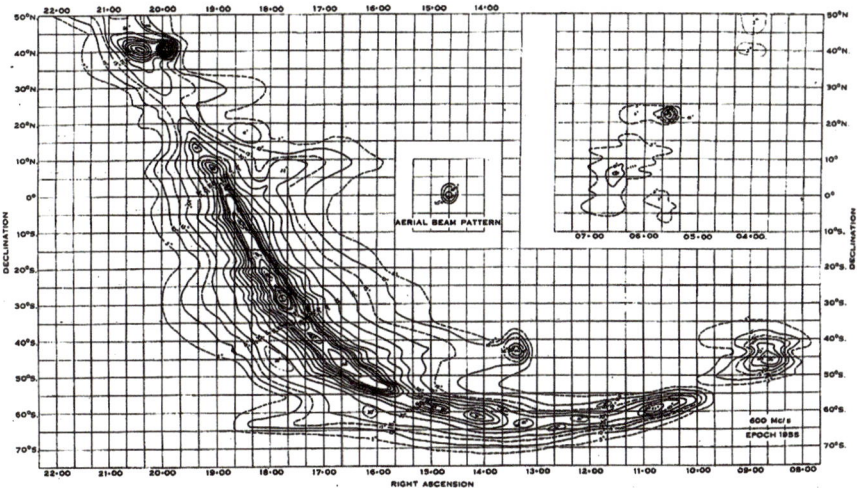

Fig. 4.37 Map of the distribution of 600 MHz emission in the general vicinity of the Galactic Plane. Unlike other figures in this chapter, the coordinates of the plot are not aligned with our Galaxy, but rather the Earth's equator (after Piddington and Trent, 1956b: 483)

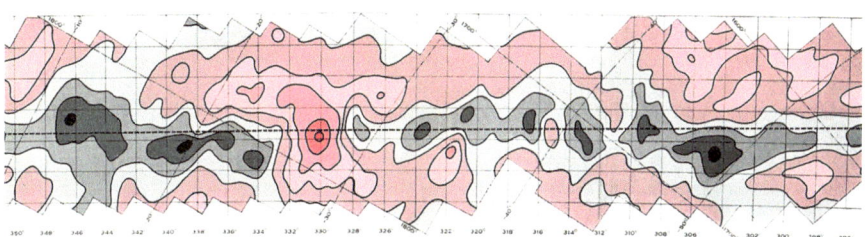

Fig. 4.38 Part of the map showing the distribution of 19.7 MHz radiation along the Galactic Plane. Dark regions indicate low signal levels along the Galactic Plane, interpreted as absorption by ionised gas (adapted from Shain, Komesaroff and Higgins, 1961: Plate 1)

Galactic emission at low frequencies. Between 1956 and 1960, Alex Shain carried out an ambitious sky survey, using this pencil-beam radio telescope in 'scanning mode' to obtain records at five declinations separated by 0.5° to 0.75° (depending upon the zenith distance of the observing region). Scans through the Galactic Plane showed distinct dips rather than the peaks that are typically exhibited at higher frequencies, indicating that "… absorption of 19.7 Mc/s radiation is occurring in a band of HII regions near the galactic plane." (Shain, 1957: 198). The 19.7 MHz map that Shain et al. (1961) published (Fig. 4.38) covers a strip of the Milky Way about 10° wide extending either side of the Galactic Centre, and it confirms the general coronal–Galactic Plane dichotomy noted by Mills et al. at 85.5 MHz.

Only with the advent of the Parkes Radio Telescope in 1961 would it be possible to complete more detailed all-sky surveys over a range of frequencies.

4.5   Distribution of Neutral Hydrogen

In 1944, in the depths of WWII, the Dutch astronomer Henk van de Hulst (1918–2000) predicted that neutral hydrogen in our Galaxy could be responsible for a radio emission line at 21 cm (or 1420 MHz) (Fig. 4.39). Although the Dutch were preparing to search for this line in 1951, it was in fact an American pair, Harold ('Doc') Ewen (1922–2015) and Edward Purcell (1912–1997), who made the discovery (Fig. 4.40). Frank Kerr (Biobox 4.5), destined to be Australia's foremost H-line researcher, has fond memories of that event. At the time, he was at Harvard College Observatory broadening his knowledge of astronomy:

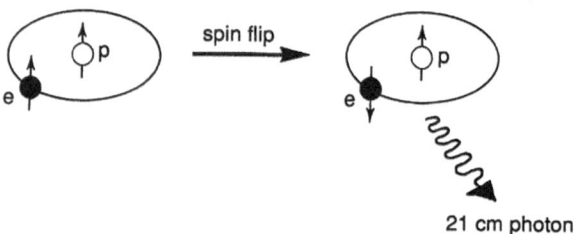

Fig. 4.39 Radio emission at 21 cm from interstellar hydrogen occurs with a flip in the spin of the electron, from parallel to the spin of the proton to the antiparallel direction. The reverse process can also occur with absorption of a 21 cm photon, which moves the hydrogen atom back to the parallel spin state at higher energy (after Robertson, 1992: 81)

Fig. 4.40 At Harvard University in July 1951 shortly after the discovery of the 21 cm hydrogen line (from left): Ed Purcell, Taffy Bowen and 'Doc' Ewen. The following year Purcell was awarded the Nobel Prize for Physics for his research on nuclear magnetic resonance (courtesy: *Life* magazine)

I happened to be there on the famous day (March 25, 1951) when H.I. ('Doc') Ewen and
E.M. Purcell first detected the 21-cm line from neutral hydrogen in interstellar material in
the Galaxy. H.C. van de Hulst, one of the pioneering Dutch radio astronomers, was also by
chance at Harvard at that time. Purcell called us all together on the morning after that over-
night discovery. Cautious scientist that he was, he wanted to see confirmation of the detection before publishing the result. He suggested that van de Hulst and I should cable our
respective institutions in Holland and Australia to report the discovery and ask whether
early confirmation would be possible. (Kerr, 1984: 137–138).

Biobox 4.5: Frank Kerr

Frank John Kerr (Fig. 4.41) was born on 8 January 1918 while his parents
were in England (as part of the war effort), and the family returned to Australia
in 1919. He studied physics and mathematics at the University of Melbourne,
graduating with BSc and MSc degrees in 1938 and 1940 respectively (Kerr,
1986). WWII halted his plans to sit for a PhD at Cambridge, but in 1962 he
was awarded a DSc by the University of Melbourne.

In 1940 Kerr joined the Radiophysics Lab and specialised in designing
radar receiver components, as well as studying radio wave propagation in the
Earth's atmosphere. The latter led Frank to a moon-bounce experiment in
1947–48, and then spending a year at Harvard University in 1950–51 to learn
more astronomy. Upon the discovery of the 21 cm spectral line of neutral
hydrogen in 1951 (coincidentally at Harvard), Frank led Australia's effort in
this branch of radio astronomy, carrying out extensive observations of the
Magellanic Clouds and our Milky Way with an 11 m dish at Potts Hill. The
Australian data were combined with a similar effort by the Dutch group in
Leiden to enable a first look at the spiral structure of the *entire* Milky Way. In

Fig. 4.41 Frank Kerr led the Australian studies of the 21 cm hydrogen line during the
1950s and early 1960s (courtesy: CRAIA)

(continued)

Biobox 4.5 (continued)

the late 1950s he was part of the team that developed the Parkes Radio Telescope, which he then used for yet more detailed mapping of hydrogen in the southern Milky Way. His research projects always emphasised international cooperation and attention to optical astronomy results, as much as to those on the radio side.

In 1966 Kerr left Sydney to join the Faculty of Astronomy at the University of Maryland. Frank continued with further studies of the dynamics and structure of the Galaxy, as well as many other projects, such as studying details of the Galactic Centre by observing its fortuitous occultations by the Moon in the late 1960s. Over two decades he trained and inspired numerous graduate students in radio astronomy (including the third author WTS of this book). Frank Kerr died in Maryland on 15 September 2000. For further details see Sullivan (1988) and Westerhout (2000).

Alexander (Lex) Muller (1923–2004) and Jan Oort (1900–1992) in Holland quickly provided confirmation, but in Australia the challenge was considerable. Although Joe Pawsey and others had been aware of van de Hulst's paper, there had been no attempt to follow it up. Bernie Mills (1976) thought that this was because

> … no one really felt strongly enough about it to put the work into what we felt, or I felt, might be a rather speculative sort of thing. There was no certainty of positive results. There were lots of other things to be done … [and the H-line] appeared rather forbidding. One knew one had to get right down to the absolute maximum theoretical sensitivity because the thing was going to be faint … it wasn't clear that it would be above the detectable limit. So it was a speculative project from that point of view.

Since no preparatory work had been done, Joe Pawsey asked both Chris Christiansen and Jim Hindman to construct H-line receivers. At first they worked independently and unbeknown to each other in adjacent instrument huts at Potts Hill located near the old Georges Heights radar antenna (see Fig. 2.42). Each RP radio astronomer made significant progress before discovering what the other was up to, and they then decided to combine their efforts.

The result was that in just six short but hectic weeks they were able to cobble together a primitive H-line receiver, which "… was held together with string and sealing wax … [and] kept going through shear will power." (*People* magazine, 1954). At a much later date Christiansen described this receiver as "… the most terrible piece of equipment I've ever seen in all my life … It was a monster …" (Christiansen, 1978). Nevertheless, he and Hindman attached the 'monster' to the ex-Georges Heights radar antenna, and succeeded in detecting the hydrogen line. Christiansen (1978) proudly remembers that "The first time I saw the line I'd [nearly] given up. I thought the gear would never work. And then I went to sleep and came back and there was a beautiful curve sitting up on the chart." (see Wendt et al., 2008 for more detail on these early 21 cm observations).

Ewen and Purcell's paper appeared in the 1 September 1951 issue of *Nature*. It was immediately followed by one penned by the Dutch radio astronomers and by a note dated 12 July, hurriedly composed by Joe Pawsey on behalf of the RP team, and cabled from Australia:

> Referring to Professor Purcell's letter of June 14 announcing the discovery of hyperfine structure of the hydrogen line in galactic radio spectrum, confirmation of this has been obtained by Christiansen and Hindman, of the Radio Physics Laboratory, Commonwealth Scientific and Industrial Research Organization, using a narrow-beam aerial. Intensity and line-width are of same order as reported, and observations near declination 20°S. show similar extent about galactic equator. (Pawsey, 1951).

Thus began Australia's assault on the H-line. For further details of the history of all aspects of 21-cm hydrogen-line research (up until 1953 and around the world) see Sullivan (2009).

During mid-1951 Christiansen and Hindman carried out exploratory H-line observations at Potts Hill, reporting their initial results in the *Australian Journal of Scientific Research* in 1952. They produced an isophote map of H-line emission extending over 270 degrees of Galactic longitude (including the Galactic Centre) along the Galactic Plane and from $l = +40°$ to $-50°$ (Fig. 4.42), and concluded that "... the source of line radiation occupies roughly the same part of the sky as does the visible Milky Way. Hence it may be assumed that the hydrogen is concentrated near the equatorial plane of the Galaxy." (Christiansen and Hindman, 1952: 454–455). The existence of double line profiles over a considerable range of Galactic longitudes was interpreted as evidence of spiral arms in our Galaxy. These same authors also prepared a simple two-page summary of their findings for an international astronomical audience, and this was published in a 1952 issue of *Observatory*. When quizzed about these conclusions in 1976, Christiansen had to admit that at the time "We were just flapping around in the dark", but the isophote plot "... was a really useful map because for many, many years radio astronomers used that map to find where the hydrogen was." (Christiansen, 1978). Further details of this early research with the ex-Georges Heights WWII experimental radar antenna are included in Orchiston and Wendt (2017).

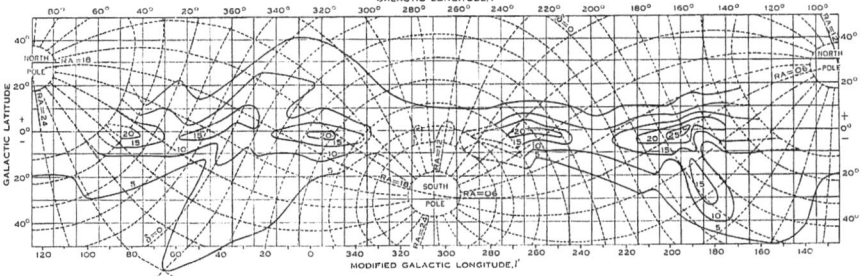

Fig. 4.42 The first published Sydney map of the distribution of neutral hydrogen in the Galaxy (after Christiansen and Hindman, 1952: 447)

Fig. 4.43 Early members of the 'H-line club' at the 1952 URSI General Assembly in Sydney (from left): Frank Kerr, Paul Wild, Jim Hindman, Doc Ewen, Alexander Muller and Chris Christiansen (courtesy: CRAIA)

As an interesting aside, in 1952 Doc Ewen came to Australia to attend the Sydney URSI Congress (Fig. 4.43) to see the RP H-line receiver for himself. Christiansen (1978) later recalled how Ewen wanted "... to see how these damn Australians did in three weeks what took [other] people eighteen months to do. And when he saw the gear he just about passed out."

In this same year Paul Wild (1952) (Fig. 2.46) published a seminal paper in the prestigious *Astrophysical Journal* on ways in which H-line investigations could be used to address some important astrophysical issues (see Wendt, 2011). As Kerr, Hindman and Robinson (1954: 297) were later to remark, the discovery of this line "... opened up new possibilities in astronomical exploration." In particular, H-line radiation could penetrate clouds of interstellar dust and reveal the distribution of hydrogen gas, while frequency displacements of the line provided invaluable information about the motion of the emitting gas. And where independent measures of distance were available, the three-dimensional structure of the gas emitting regions could be deduced. The H-line was indeed an invaluable new research tool for astronomers.

After Christiansen and Hindman completed their pioneering 1951 study, Christiansen turned his attention to solar radio emission and innovative new instrumentation (his solar grating arrays) and it was left to Hindman, Kerr and young Brian Robinson (1930–2004; Biobox 4.6) to take the H-line work further. This they did in 1953 after construction of the 11 m Potts Hill dish and what Kerr (1984: 139)

Biobox 4.6: Brian Robinson

Brian John Robinson (Fig. 4.44) was born in Melbourne on 4 November 1930 and died at Bonnells Bay, NSW, on 22 July 2004. After graduating with a BSc (First Class Honours and the University Medal) in Physics from the University of Sydney in 1952, he completed an MSc at the same university, and then joined the Radiophysics Lab in 1953. The following year, Brian moved to Trinity College, Cambridge, to study for a PhD under Jack Ratcliffe, graduating in 1958.

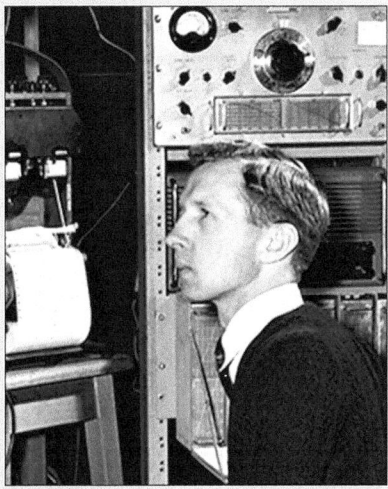

Fig. 4.44 Brian Robinson joined the hydrogen-line team at the Potts Hill field station in 1953 (courtesy: CRAIA)

Brian then spent four years at Leiden Observatory working mainly on parametric amplifiers, and brought this new technology back to the Parkes Radio Telescope in 1962. His early research was on H-line and OH-line emission, but then progressed to a study of a range of new interstellar molecules. Pulsars were another interest and one of his 1968 chart records was reproduced on the Australian $50 bank note (see Fig. 5.8). Brian was also involved in the development of new facilities, including the millimetre radio telescope at the ATNF site in Marsfield, NSW, and he played a key role in the design of the Australia Telescope Compact Array.

Brian moved quickly through the CSIRO ranks, rising to Chief Research Scientist before his retirement in 1992. He was active in the IAU and from 1975 to 1979 he chaired the Australian National Committee for Radio Science. In 1974 he was awarded the Walter Burfitt Prize by the Royal Society of New South Wales for his outstanding contributions to the field of radio physics. For further details see Whiteoak and Sim (2006).

has described as "... the world's first "multi-channel" receiver, which had all of four 40 kc/s channels!" (Fig. 2.44). They began by making the first H-line observations of extragalactic objects, in this case the Large and Small Magellanic Clouds, and published their results in the *Australian Journal of Physics* in 1954. In this important paper they showed that the neutral hydrogen extended well beyond the optical boundaries of each Magellanic Cloud (Fig. 4.45). They also showed that the total masses of neutral hydrogen in the Large and Small Clouds were about 600 and 400 million solar masses respectively, that the ratio of dust to gas in the two Clouds was very different and that both Clouds were rotating. A summary of the preliminary findings of this study was presented at a meeting of the American Astronomical Society in August 1953, and subsequently published in the *Astronomical Journal*.

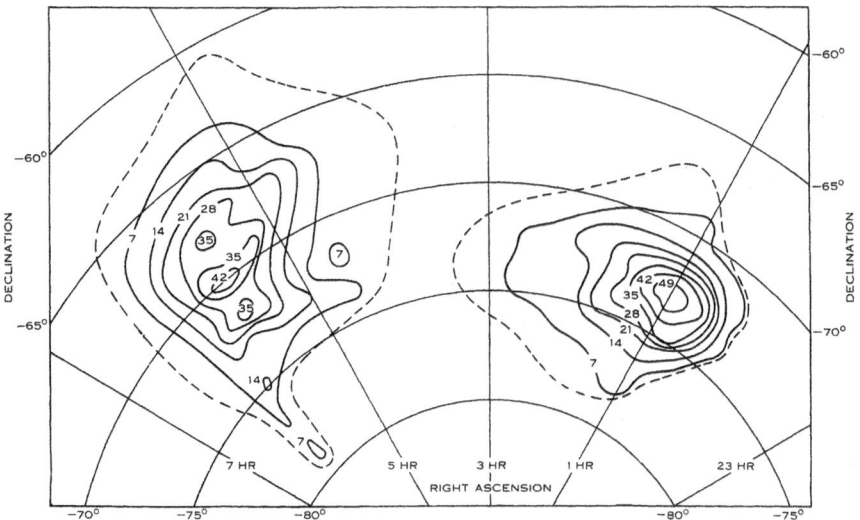

Fig. 4.45 H-line emission from the Large and Small Magellanic Clouds. The limits of detected radiation lie within the dashed lines (after Kerr, Hindman and Robinson, 1954: Fig. 1)

Kerr and the Australian-based French-born astronomer Gérard de Vaucouleurs (1918–1995) followed up these papers with a study of "... the three-dimensional distribution of gas density and rotational motion" in the Magellanic Clouds (Kerr and de Vaucouleurs, 1955: 509). After first considering the roles of the Sun's motion in the Galaxy and for Galactic rotation, they examined the evidence for rotation in each of the Clouds and found it convincing (see Fig. 4.46). Radio data supported the view that the Large Magellanic Cloud is a flattened system that is tilted by at least 65° relative to our line of sight. The tilt angle of the Small Magellanic Cloud is only about 30°, and at 1420 MHz it has "... a large prominence, or wing, extending

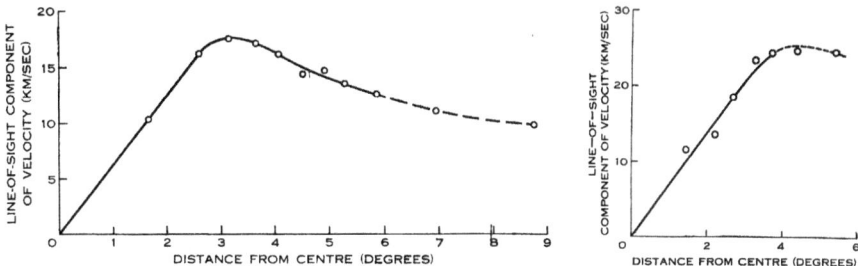

Fig. 4.46 Mean rotation curves for the Large Magellanic Cloud (left) and the Small Magellanic Cloud (after Kerr and de Vaucouleurs, 1955: 514, 515)

towards the Large Cloud." (Kerr and de Vaucouleurs, 1955: 515). This unusual feature is now known to be part of the 'Magellanic Stream'.

One of the most fascinating H-line studies carried out was a collaboration between the Sydney and Dutch groups to map the locations of the spiral arms in our Galaxy. After publication of some preliminary results of the Sydney study by Kerr and visiting American radio astronomer Martha ('Patty') Stahr Carpenter (1920–2013; Fig. 4.47), the seminal paper in *Nature* by Kerr, Hindman and Carpenter (1957) provided impressive pictorial evidence of the spiral nature of our Galaxy. As Fig. 4.48 illustrates, the Potts Hill observations provided evidence of at least four major spiral arms (the so-called Scutum-Norma, Sagittarius, Orion and Perseus Arms). The Sydney observations also provided interesting information on the distribution of hydrogen gas in the plane of the Galaxy. Meanwhile, Kerr and Dutch colleagues, Jan Oort and Gart Westerhout, took the combined Sydney and Leiden observations further in a paper published in 1958 in the *Monthly Notices of the Royal Astronomical Society*. Of particular interest is their map of the overall distribution of neutral hydrogen in the plane of the Galaxy (Fig. 4.49), which shows great irregularity and does not indicate the spiral arms as prominently as in Fig. 4.48. Nonetheless, Oort, Kerr and Westerhout (1958) concluded that our Galaxy probably belongs to Edwin Hubble's class of spiral galaxies called Sb.

Frank Kerr followed up his initial Galactic H-line survey with a detailed study of hydrogen emission from the Southern Milky Way, carried out in collaboration with RP colleagues Jim Hindman and Colin Gum (1924–1960). When coupled with a similar northern study, this was used by Gum, Kerr and Westerhout, and by Gum and Pawsey to document a new Galactic Co-ordinate System adopted by the International Astronomical Union in March 1959. As Dutch astronomer Adriaan Blaauw (1914–2010), along with Gum, Pawsey and Westerhout outlined, Australian radio astronomy played a crucial role in this important study (Blaauw et al., 1960).

Fig. 4.47 Among the notable research visitors was Martha Stahr Carpenter from Cornell University, who spent a sabbatical year at Radiophysics in 1955. At the time, women radio astronomers were a rarity. Martha's interests were in the 21 cm hydrogen line and she collaborated with Frank Kerr, Jim Hindman and Brian Robinson using the 11 m transit dish at Potts Hill (courtesy: CRAIA)

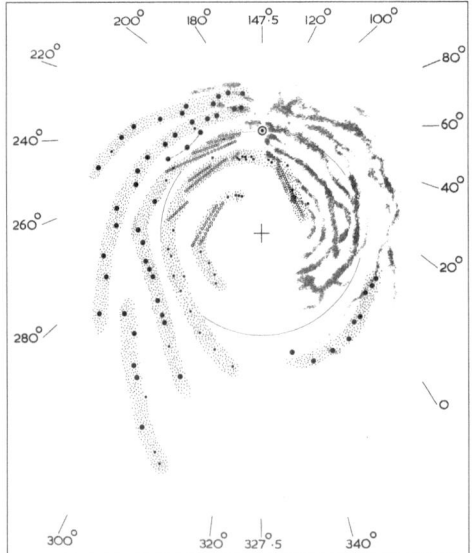

Fig. 4.48 Composite diagram showing the spiral structure of our Galaxy based on H-line observations from Potts Hill (left half) and Leiden (right half). The Galactic Centre is marked by a cross and the Sun's position and assumed circular orbit is also shown by the bull's-eye (top centre) (after Kerr, 1958: 924)

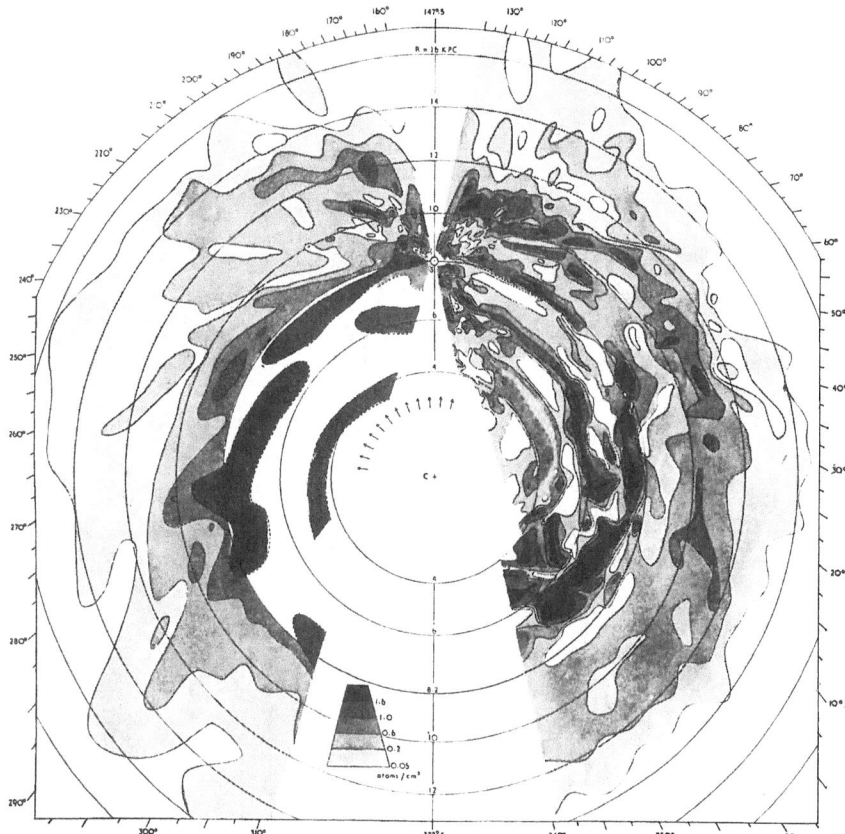

Fig. 4.49 Contour map showing the distribution of neutral hydrogen in the Galaxy, based on Australian and Dutch data (after Oort, Kerr and Westerhout, 1958: 382)

Subsequently, Kerr carried out an analysis of the differences that typified the Sydney and Leiden H-line data, and he and Westerhout penned a chapter titled "Distribution of Interstellar Hydrogen" for Volume 5 of the definitive series *Stars and Stellar Systems* (Volume 5) (Kerr and Westerhout, 1965). The Australian radio astronomers certainly had made 'good mileage' out of their pioneering Potts Hill observations.

Further H-line progress was only possible with the advent of the Murraybank field station (Fig. 2.68) in 1956, and completion of a 48-channel receiver and a 6.4 m alt-azimuth-mounted dish. The sophistication of the new receiver, and the ability to access designated regions of sky – instead of relying on the Earth's rotation as at Potts Hill – revitalised Australian H-line work. Back in 1954, Joe Pawsey foreshadowed the importance of these developments:

The crying need here [with H-line studies] is for first-class equipment. The choice of observations is still easy in this unexploited field. The current instrumental deficiency is a receiver giving an instantaneous profile. When this is satisfied there will be a bread-and-butter survey problem extending over years ... (Pawsey, 1954).

The initial plan was to acquire an 18 m dish and to use this for high-resolution H-line work, but when this antenna eventually arrived (Fig. 2.66) it ended up at Fleurs, not Murraybank, and was used for continuum studies (for details, see Orchiston and Mathewson, 2009).

Instead, Dick McGee and John Murray (1924–2020; Fig. 2.69) had to make use of the very much smaller 6.4 m antenna. In an initial test of the new dish and 48-channel receiver and its large-scale survey potential, they investigated the distribution of neutral hydrogen in the Taurus–Orion region. The new receiver allowed them to obtain an individual H-line profile in just two minutes, and to generate composite profiles by adding these. More than 3500 profiles were obtained and some 3500 square degrees of sky were surveyed. The level of emission suggested that a large single neutral hydrogen cloud or an association of connected clouds spans the Taurus–Orion region and that this is rotating as part of the general structure of the Galaxy (McGee and Murray, 1959).

Dick McGee, John Murray and research assistant Janice Milton followed up this study with a survey of the distribution of neutral hydrogen over the whole sky visible from Sydney. Meridian transit observations were made at $1°$ intervals between declinations $-90°$ and $+42°$, and more than 95,000 profiles were obtained. McGee and Murray found evidence that in general the gas was stratified parallel to the Galactic Plane, although there were some regions of excess density. Meanwhile, in our region of the Galaxy, they found that

… hydrogen is flowing away from the Sun at about 6 km/s in the direction of the galactic centre and anticentre in low and medium latitudes and is streaming in from above and below in latitudes $|b^1| = 90°$ to $|b^1| \approx 40°$ at about 6 km/s. (McGee and Murray, 1961: 278).

In a follow-up paper McGee, Murray and Milton (1963) reported on the detailed distribution of low velocity gas, and produced a composite map (Fig. 4.50) highlighting the gas concentration along the Galactic Plane. In a third and final paper in the series, McGee and Milton examined the distribution and intensity of hydrogen gas at higher radial velocities, finding that it was concentrated in a number of massive spiral arms of our Galaxy.

One of the problems encountered with the H-line receiver at Murraybank was the enormous quantity of data obtained in a relatively short time, which prompted the development of a digital data-recording system that ultimately saw extensive use with the 64 m Parkes Radio Telescope. However, this innovative system was first trialed at Murraybank during a low resolution H-line survey of the Magellanic Clouds. The digital recording and data handling system successfully converted 250 hours of observations to printed profiles in just eight hours of computer time (Hindman et al., 1963a, 1963b), but more than this, the survey reinforced the earlier finding of Kerr, Hindman and Robinson (1954) that an extensive gaseous envelope enclosed both Magellanic Clouds. The total mass of hydrogen in the two Clouds

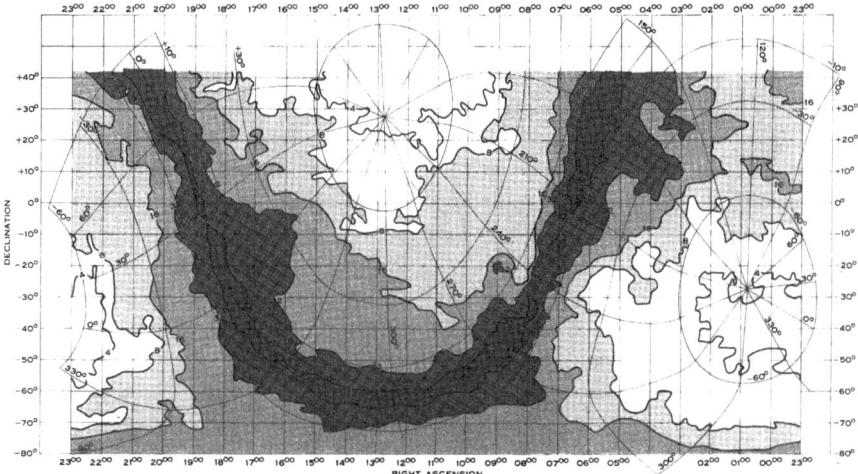

Fig. 4.50 Distribution of low velocity neutral hydrogen gas between declinations +42° and –80° (after McGee, Murray and Milton, 1963: 156)

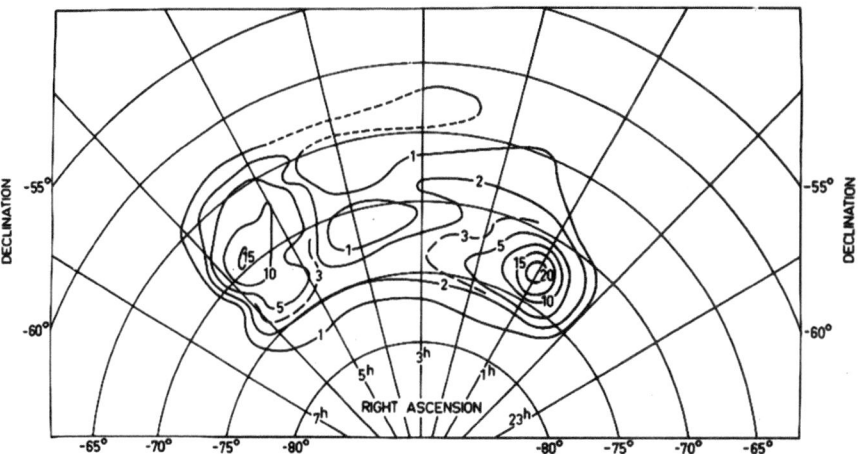

Fig. 4.51 H-line emission from the Large and Small Magellanic Clouds showing the bridge of gas linking the two systems (after Hindman, Kerr and McGee, 1963b: 572)

derived in the earlier study was also confirmed. A new discovery was the detection of a tenuous 'bridge' of hydrogen gas between the Large and Small Magellanic Clouds (see Fig. 4.51). And a further discovery, associated with the Small Magellanic Cloud, was the existence of double peaks over a wide area, suggesting the possibility that this Cloud consists of two quite separate masses of gas. With the closure of the Murraybank field station, the H-line work transferred easily and automatically to Parkes. For details of the international contribution that the Murraybank field station made to international radio astronomy, see Wendt et al. (2011a).

The 1420 MHz (21 cm) hydrogen line was the only radio line observed by the RP radio astronomers at the field stations. However, Gordon Stanley and visiting Fulbright fellow Robert Price (1929–2008) did conduct an unsuccessful search for a postulated 327 MHz deuterium line in 1954 with the 24.4 m 'hole-in-the-ground' antenna at Dover Heights (deuterium is an isotope of hydrogen with a nucleus consisting of one proton and one neutron). They had hoped that they might see this in absorption against the Sagittarius A and Centaurus A sources. At the time, they did not publish their 'null result', but when Russian colleagues reported a detection at flux levels far in excess of the upper limit they had set, they were forced into print with a rebuttal (Stanley and Price, 1956). The elusive deuterium line was eventually detected in 2007 by a group at the MIT Haystack Observatory in Massachusetts.

4.6   Radio Emission from Stars

Once the true nature of Taurus A, Centaurus A, Virgo A and other discrete sources became known in the late 1940s, the search was on for genuine 'radio stars'. Two scientists were involved in making the first detections of radio emission from stars other than the Sun: Bernard Lovell (1913–2012; Fig. 1.40) at Jodrell Bank and Bruce Slee at RP. Both began by selecting the dMe flare stars as their targets, and to Slee the reasons were obvious:

> Detection of radio emission from flare stars might lead to several important new fields of research. For example, although probably 80 percent of all main sequence stars are M dwarfs, we know little about the physical conditions in their atmospheres. Studies of radio bursts – which could be generated in these stars' coronas, if indeed they have extensive atmospheres like the Sun – would be valuable in this connection.
>
> Again, statistics of flare-star bursts may tell whether stars in general possess "star spot" cycles analogous to the 11-year sun-spot cycle. Finally, radio studies could show whether flare stars eject high-energy particles in sufficient amount to contribute appreciably to cosmic rays and, indirectly, to the radio emission of the Milky Way galaxy. (Slee, Higgins and Patston, 1963: 83).

From September 1960 to May 1962, Slee and Charlie Higgins carried out intermittent monitoring of four different flare stars: UV Ceti, Proxima Centauri, V371 Orionis and V1216 Sagittarii. Most of their observations were made at Fleurs with the N–S arms of the 85.5 MHz and 19.7 MHz Mills and Shain Crosses (Figs 2.59 and 2.63), but some observations were also carried out (at 400 and 1500 MHz) with the new Parkes Radio Telescope. Teams of amateur astronomers from Queensland, New South Wales and Victoria were enlisted to simultaneously observe during parts of the radio monitoring programs. During 1960–61 the first author WO of this book led one of these programs, using an historic 6-in Grubb refractor at Sydney Observatory (see Orchiston, 2016b: 5).

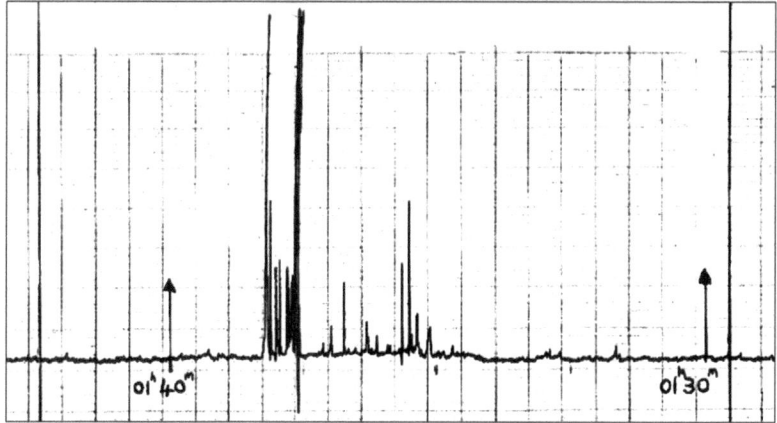

Fig. 4.52 Chart record showing the intense group of bursts observed at 19.7 MHz shortly after optical flaring of UV Ceti was recorded on 13 November 1961 (courtesy: CRAIA)

Although two strong optical flares were recorded, neither was accompanied by 85.5 MHz emission, but within a few minutes of each of three weaker flares

> … radio deflections were recorded … In all three cases … the 3.5-meter [85.5 MHz] bursts were small, but during the possible flare of UV Ceti on November 13, 1961, concurrent observations at 15 metres [19.7 MHz] revealed an extremely intense group of bursts, about four minutes after a similar but much weaker group on 3.5 metres. (Slee, Higgins and Patston, 1963: 85).

In addition, radio bursts were also recorded at 85.5 MHz on five different occasions when there was no optical monitoring.

These landmark Fleurs observations and similar results reported at much the same time by Lovell at Jodrell Bank and colleagues at the Smithsonian Astrophysical Observatory in Massachusetts (Lovell et al., 1963) marked the initiation of stellar radio astronomy. Orchiston (2004: 36–46) has documented how Bruce Slee went on to build a reputation as a world authority on radio emission from flare stars and other types of 'active stars' (e.g. see Fig. 4.52).

References

Baade, W., 1954. Letter to J. Pawsey, dated 16 February 1954, Radiophysics file A1/1/1.

Baade, W., and Minkowski, R., 1954. Identification of the radio sources in Cassiopeia, Cygnus and Puppis. *Astrophysical Journal*, 119, 206–214.

Blaauw, A., Gum, C.S., Pawsey, J.L., and Westerhout, G.H., 1960. The new I.A.U. system of galactic co-ordinates (1958 revision). *Monthly Notices of the Royal Astronomical Society*, 121, 123–131.

Bolton, J.G., 1955a. Letter to J.L. Pawsey, dated 11 July. Sullivan collection.

Bolton, J.G., 1955b. Letter to E.G. Bowen, dated 11 October. Sullivan collection.

Bolton, J.G., 1978. Taped interview with Woody Sullivan, dated 15 March.

Bolton, J.G., 1982. Radio astronomy at Dover Heights. *Proceedings of the Astronomical Society of Australia*, 4, 349–358.

Bolton, J.G., and Stanley, G.J., 1948a. Observations of the variable source of cosmic radio frequency radiation in the constellation of Cygnus. *Australian Journal of Scientific Research*, A1, 58–69.

Bolton, J.G., and Stanley, G.J., 1948b. Variable source of radio frequency radiation in the constellation of Cygnus. *Nature*, 161, 312.

Bolton, J.G., and Stanley, G.J., 1949. The position and probable identification of the source of galactic radio-frequency radiation Taurus-A. *Australian Journal of Scientific Research*, A2, 139–148.

Bolton, J.G., and Westfold, K.C., 1950. Galactic radiation at radio frequencies. I. 100 Mc/s. survey. *Australian Journal of Scientific Research*, A3, 19–33.

Bolton, J.G., Stanley, G.J., and Slee, O.B., 1949. Positions of three discrete sources of Galactic radio-frequency radiation. *Nature*, 164, 101.

Bolton, J.G., Stanley, G.J., and Slee, O.B., 1954a. Galactic radiation at radio frequencies. VIII. Discrete sources at 100 Mc/s between declinations +50° and –50°. *Australian Journal of Physics*, 7, 110–129.

Bolton, J.G., Westfold, K.C., Stanley, G.J., and Slee, O.B., 1954b. Galactic radiation at radio frequencies. VII. Discrete sources with large angular widths. *Australian Journal of Physics*, 7, 96–109.

Bowen, E.G., 1973. Taped interview with Woody Sullivan, dated 24 December.

Carter, A.W.L., 1956. Letter to R. Hanbury Brown, dated 7 December. Sullivan collection.

Christiansen, W.N., 1978. Taped interview with Woody Sullivan, dated 27 August.

Christiansen, W.N., and Hindman, J.V., 1952. A preliminary survey of 1420 Mc/s. line emission from Galactic hydrogen. *Australian Journal of Scientific Research*, A5, 437–455.

Edge, D.O., 1975. Letter to W. Sullivan, dated 7 July. Sullivan collection.

Edge, D.O., and Mulkay, M.J., 1976. *Astronomy Transformed: The Emergence of Radio Astronomy in Britain*. New York, Wiley.

Goddard, B.R., Watkinson, A., and Mills, B.Y., 1960. An interferometer for the measurement of radio source sizes. *Australian Journal of Physics*, 13, 665–675.

Hanbury Brown, R., 1957. Letter to J.L. Pawsey, dated January. Sullivan collection.

Hanbury Brown, R., and Hazard, C., 1953. A survey of 23 localized radio sources in the northern hemisphere. *Monthly Notices of the Royal Astronomical Society*, 113, 123–133.

Hey, J.S., Parsons, S.J., and Phillips, J.W., 1946. Fluctuations in cosmic radiation at radio frequencies. *Nature*, 158, 234.

Hindman, J.V., and Wade, C.M., 1959. The Eta Carina Nebula and Centaurus A near 1400 Mc/s. *Australian Journal of Physics*, 12, 258–269.

Hindman, J.V., McGee, R.X., Carter, A.W.L., Holmes, E.C.J., and Beard, M., 1963a. A low-resolution hydrogen-line survey of the Magellanic system. I. Observations and digital reduction procedures. *Australian Journal of Physics*, 16, 552–569.

Hindman, J.V., Kerr, F.J., and McGee, R.X., 1963b. A low-resolution hydrogen-line survey of the Magellanic system. II. Interpretation of results. *Australian Journal of Physics*, 16, 570–583.

Kellermann, K.I., Orchiston, W., and Slee, B., 2005. Gordon James Stanley and the early development of radio astronomy in Australia and the United States. *Publications of the Astronomical Society of Australia*, 22, 13–23.

Kerr, F.J., 1958. Sydney work on 21-cm observations of the HI Galactic disk. *Reviews of Modern Physics*, 30, 924–925.

Kerr, F.J., 1984. Early days in radio and radar astronomy in Australia. In Sullivan, W.T. (ed.). *The Early Years of Radio Astronomy*. Cambridge, Cambridge University Press. Pp. 133–145.

Kerr, F.J., 1986. Letter to Woody Sullivan, dated 12 August. Sullivan collection.

Kerr, F.J., and de Vaucouleurs, G., 1955. Rotation and other motions of the Magellanic Clouds from radio observations. *Australian Journal of Physics*, 8, 508–521.

Kerr, F.J., and Westerhout, G., 1965. Distribution of interstellar hydrogen. In Blaauw, A., and Schmidt, M. (eds), *Galactic Structure*. Chicago, Chicago University Press. Pp. 167–202 [Volume 5 in the 9-volume Series, *Stars and Stellar Systems*, (eds) G.P. Kuiper and B.M. Middlehurst].

Kerr, F.J., Hindman, J.V., and Carpenter, M.S., 1957. The large-scale structure of the Galaxy. *Nature*, 180, 677–679.

Kerr, F.J., Hindman, J.V., and Robinson, B.J., 1954. Observations of the 21 cm line from the Magellanic Clouds. *Australian Journal of Physics*, 7, 297–314.

Labrum, N.R., Krishnan, T., Payten, W.J., and Harting, E., 1964. Fan-beam observations of radio sources at 21 cm wavelength. *Australian Journal of Physics*, 17, 323–339.

Little, C.G., and Lovell, A.C.B., 1950. Origin of the fluctuations in the intensity of radio waves from Galactic sources: Jodrell Bank observations. *Nature*, 165, 423–424.

Lovell, B., Whipple, F.L., and Solomon, L.H., 1963. Radio emission from flare stars. *Nature*, 198, 228–230.

McGee, R.X., and Bolton, J.G., 1954. Probable observation of the galactic nucleus at 400 Mc./s. *Nature*, 173, 985–987.

McGee, R.X., and Murray, J.D., 1959. Neutral hydrogen gas in the Taurus–Orion region observed with a multichannel 21 cm line receiver. *Australian Journal of Physics*, 12, 127–133.

McGee, R.X., and Murray, J.D., 1961. A sky survey of neutral hydrogen at λ 21 cm. I. The general distribution and motions of the local gas. *Australian Journal of Physics*, 14, 260–278.

McGee, R.X., Murray, J.D., and Milton, J.A., 1963. A sky survey of neutral hydrogen at 21 cm. II. The detailed distribution of low velocity gas. *Australian Journal of Physics*, 16, 136–170.

McGee, R.X., Slee, O.B., and Stanley, G.J., 1955. Galactic survey at 400 Mc/s between declinations −17° and −49°. *Australian Journal of Physics*, 8, 347–367.

Mathewson, D.S., Healey, J.R., and Rome, J.M., 1962a. A radio survey of the Southern Milky Way at a frequency of 1440 Mc/s. I. The isophotes and the discrete sources. *Australian Journal of Physics*, 15, 354–368.

Mathewson, D.S., Healey, J.R., and Rome, J.M., 1962b. A radio survey of the Southern Milky Way at a frequency of 1440 Mc/s. II. The continuum emission from the Galactic disk. *Australian Journal of Physics*, 15, 369–377.

Mills, B.Y., 1952a. The distribution of the discrete sources of cosmic radio radiation. *Australian Journal of Scientific Research*, A5, 266–287.

Mills, B.Y., 1952b. The positions of six discrete sources of cosmic radiation. *Australian Journal of Scientific Research*, A5, 456–463.

Mills, B.Y., 1952c. Apparent angular sizes of discrete radio sources: Observations at Sydney. *Nature*, 170, 1063–1064.

Mills, B.Y., 1955. The observation and interpretation of radio emission from some bright galaxies. *Australian Journal of Physics*, 8, 368–389.

Mills, B.Y., 1956a. Proposed source size measurements at Fleurs. Sydney, RP. Sullivan collection.

Mills, B.Y., 1956b. The radio source near the Galactic centre. *Observatory*, 76, 65–67.

Mills, B.Y., 1959a. Galactic structure at meter wavelengths. In Bracewell, R.N. (ed.). *Paris Symposium on Radio Astronomy*. Stanford, Stanford University Press. Pp. 431–446.

Mills, B.Y., 1959b. The radio continuum radiation from the Galaxy. *Publications of the Astronomical Society of the Pacific*, 71, 267–291.

Mills, B.Y., 1960. On the identification of extragalactic radio sources. *Australian Journal of Physics*, 13, 550–577.

Mills, B.Y., 1976. Taped interview with Woody Sullivan, dated 25–26 August.

Mills, B.Y., 1985. Obituary – Little, Alec. *Proceedings of the Astronomical Society of Australia*, 6, 113.

Mills, B.Y., and Slee, O.B., 1957. A preliminary survey of radio sources in a limited region of the sky at a wavelength of 3.5 m. *Australian Journal of Physics*, 10, 162–194.

Mills, B.Y., and Thomas, A.B., 1951. Observations of the source of radio-frequency radiation in the constellation of Cygnus. *Australian Journal of Scientific Research*, A4, 158–171.

Mills, B.Y., Little, A.G., and Sheridan, K.V., 1956a. Emission nebulae as radio sources. *Australian Journal of Physics*, 9, 218–227.

Mills, B.Y., Little, A.G., and Sheridan, K.V., 1956b. Radio emission from novae and supernovae. *Australian Journal of Physics*, 9, 84–89.

Morton, D.C., 1985. The centre of our Galaxy: Is it a black hole? *Australian Physicist*, 22, 218–222.

Oort, J.H., Kerr, F.J., and Westerhout, G., 1958. The Galactic system as a spiral nebula. *Monthly Notices of the Royal Astronomical Society*, 118, 379–389.

Orchiston, W., 2004. From the solar corona to clusters of galaxies: the radio astronomy of Bruce Slee. *Publications of the Astronomical Society of Australia*, 21, 23–71.

Orchiston, W., 2005. Sixty years in radio astronomy: a tribute to Bruce Slee. *Journal of Astronomical History and Heritage*, 8, 3–10.

Orchiston, W., 2012. The Parkes 18-m antenna: A brief historical evaluation. *Journal of Astronomical History and Heritage*, 15, 96–99.

Orchiston, W., 2016a. *Exploring the History of New Zealand Astronomy: Trials, Tribulations, Telescopes and Transits*. Cham (Switzerland), Springer.

Orchiston, W., 2016b. Introduction. In Orchiston, 2016a. Pp. 1–29.

Orchiston, W., 2016c. John Bolton, Gordon Stanley, Bruce Slee and the riddle of the 'Radio Stars'. In Orchiston, 2016a, 653–671.

Orchiston, W., and Mathewson, D., 2009. Chris Christiansen and the Chris Cross. *Journal of Astronomical History and Heritage*, 12, 11–32.

Orchiston, W., and Robertson, P., 2017. The origin and development of extragalactic radio astronomy: the role of the CSIRO Division of Radiophysics Dover Heights Field Station in Sydney. *Journal of Astronomical History and Heritage*, 20, 289–312.

Orchiston, W., and Slee, B., 2002. Ingenuity and initiative in Australian radio astronomy: the Dover Heights 'hole-in-the-ground' antenna. *Journal of Astronomical History and Heritage*, 5, 21–34.

Orchiston, W., and Slee, B., 2006. Early Australian observations of historic supernova remnants at radio wavelengths. In Chen, K.-Y., et al. (eds). *Proceedings of the Fifth International Conference on Oriental Astronomy*. Chiang Mai (Thailand), University of Chiang Mai. Pp. 43–56.

Orchiston, W., and Slee, B., 2017. The early development of Australian radio astronomy: the role of the CSIRO Division of Radiophysics field stations. In Nakamura, T., and Orchiston, W. (eds), *The Emergence of Astrophysics in Asia: Opening a New Window on the Universe*. Cham (Switzerland), Springer. Pp. 497-578.

Orchiston, W., and Wendt, H., 2017. The contribution of the Georges Heights experimental radar antenna to Australian radio astronomy. *Journal of Astronomical History and Heritage*, 20, 313–340.

Orchiston, W., Nakamura, T., and Strom, R. (eds), 2011. *Highlighting the History of Astronomy in the Asia–Pacific Region: Proceedings of the ICOA-6 Conference*. New York, Springer.

Orchiston, W., Sim, H., and Robertson, P., 2016. Dr Owen Bruce Slee, 10 August 1924 – 18 August 2016. *Australian Physics*, 53, 214.

Orchiston, W., Slee, B., George, M., and Wielebinski, R., 2015. The history of early low frequency radio astronomy in Australia. 4: Kerr, Shain, Higgins and the Hornsby Valley field station near Sydney. *Journal of Astronomical History and Heritage*, 18, 285–311.

Pawsey, J.L., 1949. Letter to M. Ryle, dated 24 January. RP file A1/1/1.

Pawsey, J.L., 1951. [Note]. *Nature*, 168, 358.

Pawsey, J.L., 1954. Radio Astronomy Group – Current Program. Sydney, RP. Sullivan collection.

Pawsey, J.L., and Bracewell, R.N., 1955. *Radio Astronomy*. Oxford, Clarendon Press.

People magazine, 2 October 1954.

Piddington, J.H., 1978. Taped interview with Woody Sullivan, dated 13 March.

Piddington, J.H., and Minnett, H.C., 1951. Observations of Galactic radiation at frequencies 1210 and 3000 Mc/s. *Australian Journal of Scientific Research*, A4, 459–475.

Piddington, J.H., and Trent, G.H., 1956a. Cosmic radio sources observed at 600 Mc/s. *Australian Journal of Physics*, 9, 74–83.

Piddington, J.H., and Trent, G.H., 1956b. A survey of cosmic radio emission at 600 Mc/s. *Australian Journal of Physics*, 9, 481–493.

Robertson, P., 1992. *Beyond Southern Skies: Radio Astronomy and the Parkes Telescope*. Sydney, Cambridge University Press.

Robertson, P., 2016. John Bolton and the Nature of Discrete Radio Sources. PhD thesis, University of Southern Queensland, Queensland.

Robertson, P., 2017. *Radio Astronomer: John Bolton and a New Window on the Universe*. Sydney, NewSouth Publishing.

Robertson, P., and Bland-Hawthorn, J., 2014. Centre of the Galaxy: Sixtieth anniversary of an Australian discovery. *Australian Physics*, 51, 194–199.

Robertson, P., Cozens, G., Orchiston, W., and Slee, B., 2010. Early Australian optical and radio observations of Centaurus A. *Publications of the Astronomical Society of Australia*, 27, 402–430.

Robertson, P., Orchiston, W., and Slee, B., 2014. John Bolton and the discovery of discrete radio sources. *Journal of Astronomical History and Heritage*, 17, 283–306.

Ryle, M., Smith, F.G., and Elsmore, B., 1950. Preliminary survey of the radio stars in the northern hemisphere. *Monthly Notices of the Royal Astronomical Society*, 110, 508–523.

Scheuer, P.A.G., Slee, O.B., and Fryar, C.F., 1963. Apparatus for investigating the angular structure of radio sources. *Proceedings of the Institution of Radio Engineers Australia*, 24, 185–190.

Shain, C.A., 1957. Galactic absorption of 19.7 Mc/s radiation. *Australian Journal of Physics*, 10, 195–203.

Shain, C.A., 1958. The radio emission from Centaurus-A and Fornax-A. *Australian Journal of Physics*, 11, 517–529.

Shain, C.A., and Higgins, C.S., 1954. Observations of the general background and discrete sources of 18.3 Mc/s cosmic noise. *Australian Journal of Physics*, 7, 130–149.

Shain, C.A., Komesaroff, M.M., and Higgins, C.S., 1961. A high resolution Galactic survey at 19.7 Mc/s. *Australian Journal of Physics*, 14, 508–514.

Sheridan, K.V., 1958. An investigation of the strong radio sources in Centaurus, Fornax, and Puppis. *Australian Journal of Physics*, 11, 400–408.

Sim, H., 2013. Vale R.X. McGee (1920–2012). *ATNF News*, 74, 17–18.

Slee, O.B., 1978. Taped interview with Woody Sullivan, dated 1 March.

Slee, O.B., 1994. Some memories of the Dover Heights field station, 1946–1954. *Australian Journal of Physics*, 47, 517–534.

Slee, O.B., Higgins, C.S., and Patston, G.E., 1963. Visual and radio observations of flare stars. *Sky and Telescope*, 25, 83–86.

Smith, F.G., 1950. Origin of the fluctuations in the intensity of radio waves from Galactic sources: Cambridge observations. *Nature*, 165, 422–423.

Stanley, G.J., 1974. Taped interview with Woody Sullivan, dated 13 June.

Stanley, G.J., and Price, R., 1956. An investigation of monochromatic radio emission of deuterium from the galaxy. *Nature*, 177, 1221–1222.

Stanley, G.J., and Slee, O.B., 1950. Galactic radiation at radio frequencies. II. The discrete sources. *Australian Journal of Scientific Research*, A3, 234–250.

Sullivan, W.T., 1988. Frank Kerr and radio waves: from wartime radar to interstellar atoms. In Blitz, L., and Lockman, F.J. (eds), *The Outer Galaxy*. New York, Springer. Pp. 268–287.

Sullivan, W.T., 1990. The entry of radio astronomy into cosmology: radio stars and Martin Ryle's 2C survey. In Bertotti, R., et al. (eds). *Modern Cosmology in Retrospect*. Cambridge, Cambridge University Press. Pp. 309–330.

Sullivan, W.T., 2009. *Cosmic Noise: A History of Early Radio Astronomy*. Cambridge, Cambridge University Press.

Twiss, R.Q., Carter, A.W.L., and Little, A.G., 1960. Brightness distribution over some strong radio sources at 1427 Mc/s. *Observatory*, 80, 153–159.

Wade, C.M., 1959. The extended component of Centaurus A. *Australian Journal of Physics*, 12, 471–476.

Wendt, H., 2011. Paul Wild and his investigation of the H-line. In Orchiston, W., et al., 2011. Pp. 543–546.

Wendt, H., and Orchiston, W., 2019. The short-lived CSIRO Division of Radiophysics field station at Bankstown Aerodrome in Sydney. *Journal of Astronomical History and Heritage*, 22, 266–272.

Wendt, H., Orchiston, W., and Slee, B., 2008. W.N. Christiansen and the initial Australian investigation of the 21 cm hydrogen line. *Journal of Astronomical History and Heritage*, 11, 185–193.

Wendt, H., Orchiston, W., and Slee, B., 2011a. The contribution of the Division of Radiophysics Murraybank field station to international radio astronomy. In Orchiston, W., et al., 2011. Pp. 433–479.

Wendt, H., Orchiston, W., and Slee, B., 2011b. The contribution of the Division of Radiophysics Potts Hill field station to international radio astronomy. In Orchiston et al., 2011. Pp. 379–431.

Westerhout, G., 2000. Obituary: Frank John Kerr, 1918–2000. *Bulletin of the American Astronomical Society*, 32, 1674–1676.

Whiteoak, J.B., and Sim, H.L., 2006. Brian John Robinson 1930–2004. *Historical Records of Australian Science*, 17, 263–281.

Wild, P., 1952. The radio-frequency line spectrum of atomic hydrogen and its applications in astronomy. *Astrophysical Journal*, 115, 206–221.

Wild, P., 1978. Taped interview with Woody Sullivan, dated 3 March.

Wild, P., 1986. A tribute to Dick McGee upon his retirement, unpublished notes (Sullivan collection).

Chapter 5
Where did it all Lead?

5.1   Beyond the Doors of Radiophysics

At the end of World War II it would have been impossible to have foreseen the rapid growth of radio astronomy at the Radiophysics Lab. In the space of about five years radio astronomy not only became the dominant research program, but the RP group was easily the largest and most generously funded in the world. The two main rivals to Radiophysics were the Jodrell Bank group at the University of Manchester led by Bernard Lovell and the Cavendish Laboratory at Cambridge University led by Martin Ryle. However, both these English groups were relatively small sections within their University's physics departments and, with the post-war austerity in England, both groups operated on shoestring budgets. Around 1950, the combined budgets of the Jodrell Bank and Cambridge groups were only a small fraction of the RP radio astronomy budget (see e.g. Robertson, 1992: 131; Sullivan, 2009: 153).

And what of the United States? Even though Karl Jansky, Grote Reber and George Southworth had pioneered the field, the Americans had failed to capitalise on their early lead. It was a curious situation in view of the massive government funding of American science after the war – a level of support that saw international leadership in many fields of science shift from Europe to the United States. The potential for the Americans to add a radio astronomy string to their scientific bow had certainly been there. After the war many of the hundreds of scientists and engineers at the MIT Radiation Lab – the US equivalent of the Radiophysics Lab – returned to academic posts but, oddly, no major radio astronomy group emerged in any of the universities. A number of important, yet isolated discoveries in radio astronomy were made by physicists, most notably the discovery of the 21 cm hydrogen line by Ed Purcell and 'Doc' Ewen at Harvard (see previous chapter), but some of these discoveries never developed beyond interesting sidelines (Sullivan, 1988: 336–338). Perhaps the most significant radio astronomy program carried out in the immediate post-war years was by a group at the Naval Research Laboratory in Washington led by John Hagen and Fred Haddock. The Washington group

© The Author(s) 2021
W. Orchiston et al., *Golden Years of Australian Radio Astronomy*,
Historical & Cultural Astronomy, https://doi.org/10.1007/978-3-319-91843-3_5

constructed a 15 m parabolic dish, at the time the largest in the world, built specifi-cally to operate at wavelengths as short as 1 cm. In addition, from 1948 and during the 1950s a small group of astronomers and electrical engineers at Cornell University developed radio astronomy, and carried out solar, Galactic and extragalactic obser-vations (see Campbell, 2019). As Sullivan (2009: 200–211) has pointed out, Cornell and MIT were the first US universities to mount post-war research programs in radio astronomy.

Mention should also be made of Grote Reber. After joining the National Bureau of Standards in Washington in 1947, he began designing a new telescope with a structure similar to the successful dish at his Wheaton home, but with a projected diameter of 67 m. Reber lobbied hard in Washington circles to get support for the project, but without success. In 1951 he resigned his position and returned to radio astronomy the way he had started, all by himself, first in Hawaii and then in Tasmania. Paradoxically, Reber believed the problem was not a shortage of funds, it was with the American scientists:

> The situation was different in the United States. There was tremendous wealth and we could embark on huge projects such as nuclear energy. After the war that's where the real empires were made in science. Instead of tens of thousands of dollars, there were tens of millions to spend. The scientific community was totally mesmerised by these sums of money available to build big machines. Unlike Australia and England, the American scientists simply failed to see the potential in radio astronomy. (Robertson, 1986).

There were two significant events that helped to kick start radio astronomy in the United States. The first was the decision by the newly formed National Science Foundation to sponsor a nationwide symposium on radio astronomy. The sympo-sium was held in Washington in January 1954 and brought together most American researchers in the field, many of whom were concerned that the US had fallen behind in this exciting new field. Taffy Bowen and Bernie Mills attended and 'flew the flag' for Australia. In 1952 there were only six American institutions active in radio astronomy, but by the time of the symposium the number had grown to ten, with several more still deciding whether to take the plunge. With a couple of excep-tions all these institutions were universities, and a growing number of them were developing graduate programs in radio astronomy. A handful of Americans had already graduated with PhDs in the subject and there were a lot more on the way. The American giant was awakening. The most important outcome of the Washington meeting was the decision to support the idea of a National Radio Astronomy Observatory which would build and operate a number of large radio telescopes. Rather than individual university departments attempting to build large telescopes for their own use, the federally funded NRAO would do it for them and make the instruments available to the universities. This was a radical idea. Although progress at the national facility proved slower than expected, by the 1970s NRAO had become the powerhouse of world radio astronomy (for a comprehensive history of the NRAO, see Kellermann, Bouton and Brandt, 2020).

The second boost to American radio astronomy came in 1954 with the announcement by the Californian Institute of Technology (Caltech) of its plans to build the world's finest observatory for radio astronomy. The Caltech astronomers in Pasadena already had access to the world's finest collection of optical telescopes on Mt Wilson and Palomar Mountain, so the new radio observatory would be a natural complement. Early in 1955 Caltech recruited John Bolton and Gordon Stanley from Australia to design and build the new observatory at Owens Valley in northern California (Cohen, 1994; Robertson, 2017: Chapter 7). Their departure led to the closure of Dover Heights, which for almost ten years had been one of the most important and successful of the RP field stations.

In addition to the NRAO at Green Bank, West Virginia, and the Caltech group at Owens Valley, a number of other university centres underwent rapid development in the late 1950s including Cornell, Harvard, Illinois, Michigan, Ohio State and Stanford. Most of these groups were in advanced stages of planning a major new instrument for the 1960s. In fact the main problem in the United States was not a shortage of new instruments (or of course funds), but the lack of suitably qualified radio astronomers. One study has estimated that, around 1960, for every American-born radio astronomer there was at least one other from either Radiophysics or Jodrell Bank active in the United States (Edge and Mulkay, 1976). As Hanbury Brown at Jodrell Bank half-joked to Joe Pawsey: "You have the best radio astronomy group in the world – the trouble with being a success is that everyone is after your staff! I do hope you are managing to manufacture some more. We have a good deal of the same trouble here – we lose them to the USA and can never get them back." (Hanbury Brown, 1960).

With the Radiophysics group as the world leader at this time, Australian radio astronomers were naturally in great demand in the United States. Although John Bolton returned to Australia to take charge of the Parkes telescope (see below), Gordon Stanley stayed on and became the Director of the Owens Valley Radio Observatory (Kellermann et al., 2005). Others such as Ron Bracewell (Stanford University) and Frank Kerr (University of Maryland) took up permanent academic posts and founded their own radio astronomy groups (e.g., see Bracewell, 2005). Other RP staff went for extended visits on fellowships and helped to accelerate the growth of radio astronomy in the United States, including Bernie Mills (1954 Caltech), Jim Roberts (1958–60 Owens Valley), Kevin Westfold (1958 Owens Valley), Brian Cooper (1959 Harvard), Don Yabsley (1960–62 Cornell), Alec Little (1963 Stanford) and Dick McGee (1963 Michigan). It would be wrong to label this migration of expertise to the United States as another alarming case of the 'brain-drain' which has adversely affected many other branches of Australian science; for example, the exodus over decades of many of Australia's best young physicists to overseas centres such as the Cavendish Laboratory in Cambridge. On the contrary, here was the first and probably only occasion when a major new science pioneered in Australia played a leading part in promoting the growth of its counterpart in the United States.

5.2   Big Science comes to Australia

Perhaps the single most important event at Radiophysics during the 1950s was the decision to build a Giant Radio Telescope. The decision transformed the future development of radio astronomy in Australia and led to a major upheaval of personnel at Radiophysics. Until then no university department or government research lab in Australia had anywhere near the resources to undertake such an expensive and technologically advanced project. The construction of the Parkes Radio Telescope brought what has been termed 'Big Science' to Australia. From the outset the Parkes dish was an international project. Approximately half the funds came from American philanthropic foundations, the design was carried out by a British engineering firm and, after an international tender, the construction was contracted to a German steel company.

During the early 1950s a debate arose within the RP group about where this rapid growth in radio astronomy was heading. Should the Lab continue with a number of small groups, each working independently, or should it begin to focus on one or two large projects that would bring these small groups together? Joe Pawsey and a number of senior staff were in favour of continuing with the status quo, an approach that had brought world acclaim to the Lab. Why make changes to something that was working so well? This approach was entirely consistent with Pawsey's own background where he completed his PhD at the famous Cavendish Laboratory. He became an advocate of the so-called 'string and sealing wax' approach to research at the Cavendish, where innovative solutions to problems could be found using relatively simple and inexpensive equipment.

However, this was not the view of Taffy Bowen, Chief of the Radiophysics Lab (see Fig. 5.1). While Bowen directed the Rainmaking Group and left the day-to-day management of the Radio Astronomy Group to Pawsey, he took a very keen interest in the development of both groups. Bowen and several senior radio astronomers believed that radio astronomy would soon develop in the same way as conventional optical astronomy, where the most important discoveries were made with the largest and most powerful optical telescopes. The event that convinced Bowen that a large general purpose telescope was the way of the future came in 1951 when Bernard Lovell's group at Jodrell Bank announced plans to build a giant parabolic dish. Bowen feared that the new telescope would scoop the pool of discoveries and that Radiophysics would soon lose its position as a world leader in radio astronomy. Bowen believed that to avoid being left behind, Radiophysics would have to build an instrument at least as good as, if not better than, the Jodrell Bank giant. Although the approach of small semi-autonomous groups at Radiophysics continued until the late 1950s, Bowen's vision of a Giant Radio Telescope (GRT) eventually prevailed.

The immediate problem Bowen faced was where would the money come from to build a GRT? Radio astronomy was only one of several research projects in the Lab and as Chief he had to work within a fixed annual budget. Similarly, the Radiophysics Lab was only one of a dozen Divisions within CSIRO, all clamouring for more funding to support their expanding post-war research programs. At Jodrell Bank,

Fig. 5.1 Taffy Bowen was the driving force behind the Parkes Radio Telescope. He believed that a giant all-purpose dish would be the best way for the Radiophysics group to stay at the forefront of international radio astronomy [courtesy: CSIRO Radio Astronomy Image Archive (CRAIA)]

Bernard Lovell had received a large grant from the philanthropic Nuffield Foundation to partly fund his large telescope. Lovell had then persuaded the UK's Department of Scientific & Industrial Research to fund the remainder of the projected costs. Bowen realised he would need to explore a similar path, but was faced with the reality that there were no comparable philanthropic foundations in Australia to support scientific research, especially in a new and relatively obscure area such as radio astronomy.

During WWII Bowen had spent several years shuttling back and forth between England and the United States, where he acted as the principal liaison officer coordinating the development of radar in the two countries. Bowen spent much of his time at the MIT Rad Lab near Boston where he got to know many powerful figures in American science, including some who had considerable influence within the network of large US philanthropic organisations. He would draw on this 'old-boy network' to provide the funds to build a GRT in Australia. The breakthrough occurred when Bowen learnt that the Carnegie Corporation administered a special fund which had been established specifically to support projects within Britain and her Commonwealth countries. Bowen was informed that the trustees of the Fund had decided that, rather than support a range of small projects, they wanted to make one single large grant. Bowen was in the right place at the right time. In May 1954, the Carnegie Corporation announced that it would provide $US250,000 towards the partial funding of a GRT in Australia. With the Carnegie promise, Bowen's next challenge was to seek funding from the Australian Government. With the support of CSIRO head office, the request went to Richard Casey, the Minister-in-Charge of CSIRO, who in turn discussed the GRT project with Prime Minister Robert Menzies. Although Menzies agreed to match the Carnegie grant with government funds, he insisted that at least one half of the total costs of the project were to be raised from private sources.

By early 1955 it was clear to Bowen that the funds promised would be nowhere near enough to build a GRT to rival the telescope already under construction at Jodrell Bank. Once again he flew to the US to visit a number of other philanthropic foundations. At the Sloan and Ford Foundations he was informed that funding scientific research in another country halfway round the world was not part of the charter of either organisation. His reception at the Rockefeller Foundation was, however, far more positive, helped no doubt by his wartime colleague Lee DuBridge who was an influential member of the Rockefeller Board of Trustees. In December 1955, the Rockefeller Foundation announced that, similar to Carnegie, it too would grant $US250,000 towards the cost of an Australian GRT. With the Australian Government's pledge to match the US funding, just over $US1 million was now in hand, enough for the project to proceed.

Immediately after the announcement of the Carnegie grant in May 1954, Bowen established a GRT Planning Committee consisting of senior Radiophysics staff and external engineering experts. The committee's brief was to invite organisations to submit design concepts and then to assess their viability. Various designs were received from a number of engineering firms in Australia and overseas, including one from the renowned American inventor Buckminster Fuller. For all the ingenuity of these designs (see Fig. 1.20), the real breakthrough in the design of the GRT came about by accident. During a trip to London in 1955 Bowen was introduced to Barnes Wallis, the leading British aeronautical engineer (Fig. 5.2). He was the designer of aircraft such as the famous Wellington bomber, which became the RAF's workhorse during WWII. Over lunch one day Bowen discussed with Wallis the GRT planned for Australia. Wallis immediately came up with a few ideas and agreed to work on a design concept. The plan was then to use Wallis' ideas and carry out the detailed design of the telescope in Australia, but it became clear that there were no local firms with sufficient experience or expertise. After several engineering firms in the UK were invited to bid, Freeman Fox & Partners in London were chosen for the detailed design. The firm specialised in the design of bridges and in fact it was the firm's founder, Sir Ralph Freeman, who designed the famous Sydney Harbour Bridge in Australia.

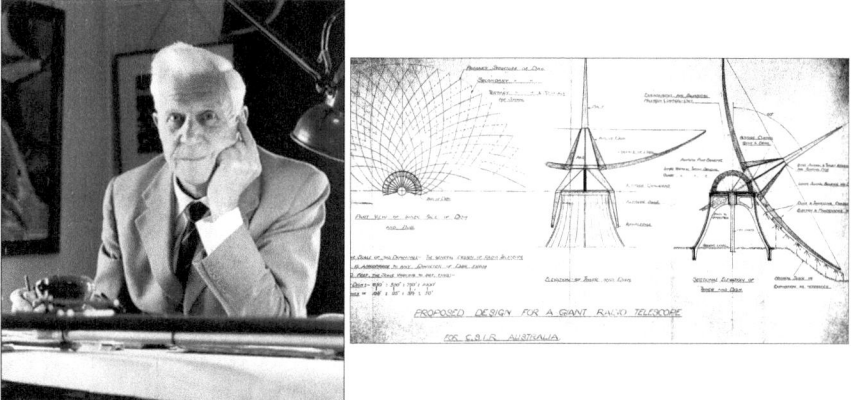

Fig. 5.2 Britain's leading aeronautical engineer Barnes Wallis carried out the initial design study for the Australian Giant Radio Telescope and most of his innovative ideas were incorporated into the final design. Based on the funds available, the engineering firm of Freeman Fox chose a dish diameter of 64 m (210 ft). Although this was less than the 76 m diameter of the Jodrell Bank dish, the greater surface and pointing accuracies made the Australian dish a superior instrument (courtesy: Vickers London and CRAIA; all rights reserved)

Throughout the planning of the GRT, Bowen and the Radiophysics group followed the progress of the telescope at Jodrell Bank with great interest. Early in the project Lovell made a number of design changes in an attempt to improve its performance. The major change, following the discovery of the hydrogen line in 1951, was the decision to operate down to a 21 cm wavelength, compared with the previous metre-wavelength lower limit. This meant finer, and heavier, mesh surface panels and the need to strengthen the support structure. The various changes led to delays in the project, contractual problems with suppliers and, worst of all, a massive cost blow out (Lovell, 1968). The Australians learnt a lesson from Lovell's troubles. Jodrell Bank had been designed in relative haste and took over five years to build. In contrast, Parkes took three years to design, but only two years to build.

The selection of the site near Parkes in central New South Wales for the new telescope took the best part of four years to settle. Preliminary discussions on a suitable site began in mid-1954 and by early 1956 a list of over thirty possible locations had been compiled. Several technical requirements were taken into consideration in shortlisting a site. The ideal location would need to be geologically stable to provide a solid foundation capable of supporting a structure weighing close to 2000 tonnes. The site would also need to have a mild climate free from ice and snow, with a low average windspeed all year round. Above all, the site had to offer a very low level of radio interference. To add to his troubles at Jodrell Bank, Lovell had been drawn into a long battle trying to persuade a local electricity authority to reroute a new high-voltage line away from the site.

While Cheshire – where Jodrell Bank was located – proved an uncomfortably noisy part of England, it was clear that a country as large and as sparsely populated as Australia would have no shortage of sites free of radio noise, and which would also meet the other geological and meteorological requirements. Although for a time a site near Canberra was in the mix, the issue settled down into a two-way contest – a site close to Sydney, or one 'over the mountains' well to the west of Sydney. A site close to Sydney had the strong advantage of convenience. From the beginning, most of the RP field stations had been no more than an hours' drive away. Staff had grown accustomed to a pattern of frequent, often daily visits to and from home to carry out their observing programs. For these reasons most of the sites examined initially fell within a comfortable radius of Sydney, the best candidate to emerge being an area known as Cliffvale, at the foot of the Blue Mountains, 80 km south–west of Sydney.

While the Cliffvale site became the favoured option, an extensive search was also made during 1957 for a location further west of Sydney, and west of the mountain range known as the Great Dividing Range. Large areas in the sparsely populated region were inspected, with a site near Parkes, about 400 km west of Sydney, chosen as the best of these 'over the mountains' candidates. Situated in the Goobang Valley, 20 km to the north of the Parkes township, this site was particularly well shielded from radio interference by a surrounding ring of hills. Tests at both sites showed that Parkes had much lower noise levels than Cliffvale. More importantly, with the rapid expansion of Sydney's western suburbs, the quality of the Cliffvale site would only deteriorate over the long term. The decision by the RP group in favour of Parkes was unanimous. Construction began at the site early in 1959 (Fig. 5.3).

Fig. 5.3 The construction of the Parkes dish during 1959–1961 was carried out by the German steelmaker MAN. Most of the components of the telescope were cast at the MAN plant near Frankfurt and then shipped to Sydney (courtesy: CRAIA)

5.3   Changing of the Guard

The inauguration of the Parkes Radio Telescope in October 1961 marked an important stage in the development of science in Australia. On a scientific level the telescope provided Australian astronomers with the most powerful and versatile instrument of its type in the world, one which immediately produced a stream of significant and, at times, fundamental discoveries. On another level the Parkes telescope also had a major impact in shaping the way astronomy developed in Australia. In contrast to the 1950s, when small teams at Radiophysics built and had exclusive use of their own telescopes, Parkes would operate as an observatory and have more in common with the great optical telescopes of the world. A specialist group of technicians would look after the routine maintenance and operation of the telescope and radio astronomers would now have to compete with their peers for observing time on the new instrument. A new breed of radio astronomer emerged and also a new way of doing radio astronomy. In this respect the Parkes telescope marked the arrival of radio astronomy in Australia as a mature scientific discipline. In some respects too, it marked the end of the most innovative and interesting period. And for those of us personally involved in the pre-Parkes era (including the first author WO of this book), it was also a particularly exciting era to be involved in radio astronomy, when we felt genuine affection for the field stations (e.g., see Orchiston and Slee, 2017: 498–499).

The Parkes telescope also had major repercussions on the broader development of astronomy in Australia. With the massive investment of staff and funds in Parkes, only limited resources remained to support the other research programs which had flourished at Radiophysics in the 1950s. In this section we see how this heavy commitment to Parkes sparked off a fierce struggle for these remaining resources that, in turn, precipitated the departure of several leading members of the RP group to continue their careers in astronomy elsewhere.

Even before the first blueprint had been drafted for the Parkes dish, the RP group realised it faced a major reorganisation of its resources. As we have noted earlier, radio astronomy began as an unlisted item in the 1946 budget and then in a few short years had grown to become the dominant research area at RP. This rapid expansion had been funded essentially by winding back on other RP research areas such as air navigation and computing. However, this changed after a crucial meeting in October 1956 between senior RP staff and the CSIRO Executive in Melbourne. The meeting agreed that resources would be provided to the Parkes dish not by further cuts to the other RP research areas, but by winding back other parts of the radio astronomy program (Robertson, 1992: 131). For the Parkes group the future was clear cut. With a superlative instrument on the way, the main tasks ahead were to ensure that the telescope was brought into operation as quickly as possible and then to start working through the wide variety of possible observing programs.

In sharp contrast, for the radio astronomers who would not be directly involved in the Parkes project, the future seemed far less certain. As we have seen earlier, Radiophysics operated a number of instruments at this time – the solar radiospectrograph at Dapto, two small dishes for 21 cm work at Potts Hill and Murraybank, and the Mills Cross, the Shain Cross and the Chris Cross at the Fleurs field station. Each of these instruments had been brought into operation during the period 1952–57 and, though none was ready for the scrap-heap, serious thought had to be given to a new generation of telescopes over the decade ahead. Any new instrument would inevitably be larger and more expensive than its predecessor, and clearly not all of the research programs of the 1950s could continue to flourish. The Parkes telescope now meant that only about half the staff and financial resources were available to stretch across all these programs. Unavoidably, cutbacks were the order of the day for the remainder of the radio astronomy group.

Two proposals for new instruments emerged as the major contenders for the 1960s. One was a metre-wavelength Radioheliograph devised by Paul Wild to supersede the Dapto spectrograph as the principal instrument for studying solar radio bursts. This would provide real-time TV-like images of the radio Sun and show the motion and evolution of bursts. The other proposal by Bernie Mills called for a SuperCross, a larger version of the instrument built at the Fleurs field station in 1954 to continue the highly successful program on the detection and cataloguing of radio sources. Both proposals had outstanding scientific merit. And yet, because of the limited resources available there was little possibility of proceeding with both. The issue would be one resolved on non-scientific grounds and here, on two counts, Wild's proposal gained the upper hand. The Dapto spectrograph had been built in 1952 and since then the Cosmic Group had constructed both the Mills and the Shain Crosses. For what it was worth, fair play suggested that the Solar Group should have the next turn. Another consideration was the question of diversification. The SuperCross would continue a research program separate and complementary to the Parkes telescope, but nevertheless it would mean a total commitment by the RP group to Galactic and extragalactic radio astronomy. Inevitably the Solar Group, by then widely recognised as the world leader, would run down and eventually be disbanded. A vote for the Radioheliograph would be a vote for diversification of the group's activities. For these reasons, at a meeting in August 1959 the radio astronomy group decided in favour of Wild's proposal.

Wild had initially proposed an instrument in the form of a cross and, only later, devised an array consisting of 96 identical aerials positioned at equal distances around a large circle. The plan had been to build the instrument at Parkes but the sheer size of the circular array, 3 km in diameter, proved too large for the 170 hectare site. This led to the decision to head north to less expensive farming country and to the eventual selection of the site at Culgoora, near Narrabri in northern New South Wales. Funding for the Radioheliograph was obtained from the US Ford Foundation in two instalments totalling $US630,000 – approximately the same amount as the combined Carnegie and Rockefeller grants made earlier towards the Parkes dish. In contrast to Parkes and the engineering challenges it presented, the simple form for each of the 96 aerials enabled the entire Radioheliograph to be built by RP staff (Fig. 5.4). After eight years of planning and construction, the Radioheliograph recorded its first signals in August 1967 (Frater et al., 2017; Chapter 5).

Fig. 5.4 Paul Wild shortly after the completion of the Radioheliograph. Originally the plan had been to build the instrument on the Parkes site, but it needed a much larger area. A new site was found at Culgoora in northern NSW which, in 1988, also became the location of the Australia Telescope Compact Array (see Section 5.3). (right) Four of the 96 antennas equally spaced around a circle 3 km in diameter (courtesy: CRAIA)

The decision in August 1959 to proceed with the Radioheliograph at the expense of the SuperCross meant that Bernie Mills had to quickly assess his future. He did not have to look far. In May 1960 the offer of a position came from charismatic Canadian-born Harry Messel in the School of Physics at the University of Sydney, virtually next door to the RP Lab (Mills, 2006; Frater et al., 2017: Chapter 3). After his appointment as Head of the School, Messel had begun a vigorous campaign to recruit new staff with the aim of transforming what was then a scientifically moribund department into a research centre of world class. To broaden the scope of the School, Messel announced he would establish a new centre for radio astronomy with Mills as his star recruit. Messel's timing was perfect. The National Science Foundation (NSF) in Washington announced that it would, for the first time, consider making grants to scientists outside the United States. Mills submitted a detailed funding application and, in 1962, the NSF announced that it would fund the SuperCross with a sum of $US746,000 spread over five years. This grant was the largest made to any project outside the United States by the NSF, a body which until then had focused almost exclusively on funding research in American universities and laboratories. Following the Parkes dish and the Radioheliograph, the SuperCross became the third major Australian radio telescope to be partly or wholly funded by the United States (Fig. 5.5).

The SuperCross is notable for being the project that finally broke the near monopoly held by Radiophysics on radio astronomy in Australia. Some radio astronomy research areas had been initiated in Australian universities, but these programs were not genuine competitors to Radiophysics. As undoubtedly the most

Fig. 5.5   A view along the east–west arm of the SuperCross at Hoskinstown, near Canberra. Bernie Mills with Australian Prime Minister Sir Robert Menzies at the inauguration of the SuperCross in November 1965 (courtesy: University of Sydney Archives)

outstanding example, a group led by Bill Ellis at the University of Tasmania in Hobart carried out observations of radio emission at low frequencies between about 1 and 30 MHz, a program boosted by the arrival of the American pioneer Grote Reber in 1954 (e.g. see George et al., 2015, 2017). Reber had been attracted by an unusual hole or 'window' in the ionosphere that makes Tasmania a particularly good location for radio astronomy at very low frequencies. Characteristically, using his own resources, Reber constructed a radio telescope at Bothwell, north of Hobart, a vast array of wooden poles and wire (see Fig. 1.42). In terms of its physical dimensions, it was the world's largest telescope – the first 'Square Kilometre Array'. Working largely alone, Reber carried out a survey of radio emission from the Milky Way at 2 MHz, a frequency much lower than anything attempted at Radiophysics. While the resources available to Reber and the University of Tasmania group were small compared to the RP radio astronomers, Tasmania quickly established itself as the world leader in very low frequency radio astronomy. Reber and Ellis' group were never viewed as competitors by RP – rather their programs complemented those of RP and helped maintain Australia's world supremacy in radio astronomy during the 1950s and through into the 1970s.

As we have seen in previous chapters, a major part of the radio astronomy program at Radiophysics in the 1950s had used instruments based on variations of the same principle. The Mills Cross had been used for surveys of Galactic and extragalactic sources, the Shain Cross for a survey of the Galaxy at low frequency and the Chris Cross for studies of the Sun. In 1960, in a period of just a few months, each of these three programs came to an abrupt halt. In February Alex Shain died, aged only 38, and the RP group lost one of its foundation members. Three months later, both Mills and Chris Christiansen joined the University of Sydney. In contrast to Mills, however, who left because of much better research prospects, Christiansen chose to leave because of his fundamental objection to the direction radio astronomy had taken at Radiophysics (Frater et al., 2017: Chapter 4). Paradoxically, Christiansen had contributed solidly to the planning of the Parkes telescope and he was the one who found the Goobang Valley site near Parkes. And yet, throughout the project he was the telescope's most vocal critic. Christiansen had achieved outstanding success in developing new radio telescopes, beginning with the grating interferometers at Potts Hill in the early 1950s. Given the brilliant record by Radiophysics in devising and building new types of telescopes, Christiansen believed the new Parkes dish to be too conventional and, indeed, too old fashioned. He mockingly referred to it as 'the last of the Windjammers' (Christiansen, 1990).

Christiansen's outspokenness put him on a collision course with his chief. The matter came to a head when Taffy Bowen learnt that Christiansen had been openly critical of Bowen's ability as a scientist. Christiansen had no option but to resign. Earlier in 1960, Christiansen had been invited to apply for the Chair in Electrical Engineering at the University of Sydney, but had turned down the offer. However, in view of his clash with Bowen and also the fact that the engineering department would be closely associated with the new radio astronomy centre, Christiansen put in a late application which was accepted immediately.

Christiansen then faced the daunting task of starting his new university position with no radio telescope and no ready source of funds. He then learnt of a plan to demolish his abandoned Chris Cross at the Fleurs field station. Following some swift negotiations, CSIRO agreed to transfer ownership of the Fleurs site to the University of Sydney. Over the next decade Christiansen revamped the old cross by adding several medium-size parabolic dishes to the array to form the Fleurs Synthesis Telescope (see Section 5.4 below). After the Parkes dish, the Radioheliograph and the SuperCross, the Fleurs Synthesis Telescope became the fourth major Australian radio telescope built in the 1960s. The Fleurs instrument would take much longer to complete than the other three, but it could be funded within the university budget and did not rely on American philanthropy.

Both Christiansen and Mills, then in their forties at the time of their controversial departure from Radiophysics, were able to re-establish themselves and carve out successful new careers in radio astronomy. For Joe Pawsey, the future did not hold the same good fortune. Since the beginning of the radio astronomy group in 1945, Pawsey had provided leadership and cohesiveness to a loose federation of small research teams; his wisdom, intellect, modesty and international stature as a scientist had won him the admiration and, in most cases, fierce loyalty of his staff. By 1960 circumstances had changed radically. With the formation of the two largely independent Parkes and Culgoora groups, Pawsey's position no longer had the same relevance.

The changing of the guard had been signalled by the appointment of John Bolton to take charge of the Parkes Radio Telescope, a post which later was given the title of Director of the Australian National Radio Observatory (ANRAO). Bolton had outstanding credentials for the position, not least his illustrious start in the late 1940s with the discovery and identification of the first discrete radio sources (see

Chapter 4). As noted above, in 1955 he and Gordon Stanley were recruited by Caltech to be the founders of the Owens Valley Radio Observatory in northern California, the first major centre for radio astronomy in the United States. A further point in Bolton's favour was that, from the outset, he had been a strong supporter of building the giant Australian dish. His support for the project, combined with a special rapport he had developed with Taffy Bowen, made him the leading candidate for the director's position. In June 1960 Bolton informed Caltech that he would be returning to Australia and, by the time the position was advertised several months later, the outcome was a foregone conclusion.

Bolton's responsibilities were to be wide ranging. Aside from pursuing his own research interests, he would oversee the allotment of observing time on the telescope, supervise a procession of postgraduate students from Australian and overseas universities, and vet all research papers before they went out for publication – duties which in practice made him leader of the Cosmic Radio Astronomy Group. In terms of professional classifications within CSIRO, Bolton was appointed at the level of Chief Research Scientist. Earlier, Paul Wild had also been elevated to the same classification after being offered a new chair in astronomy at Cornell University, at more than three times his CSIRO salary. Although CSIRO could not match the American offer in monetary terms, the promise of promotion and support to build the Radioheliograph had persuaded Wild to stay in Australia. In practice, Wild was now leader of the Solar Radio Astronomy Group. Pawsey too was classified as a Chief Research Scientist. Although he was not outranked by his two former protégés, his position had basically become redundant.

Pawsey was not to be stranded. In December 1961 he accepted an offer to succeed Otto Struve as Director of the National Radio Astronomy Observatory, recognised as the top-ranking post in radio astronomy in the United States (Fig. 5.6). Since its foundation four years earlier, the NRAO had worked towards its goal of establishing a national facility with state-of-the-art equipment, shared by radio astronomy groups across the country, as well as its own permanent research staff. A 26 m telescope had been completed in 1958 at the Green Bank site in West Virginia and a 91 m dish, steerable in elevation only, was nearing completion. In March 1962 Pawsey arrived at Green Bank for an initial six-week visit, but a few days later suffered paralysis of the left arm and leg. After several weeks in hospital he was operated on for the removal of a brain tumour. He recovered sufficiently to return to Sydney and he was able to visit the Lab a few times a week. However, his condition continued to deteriorate and he went into hospital for the last time in October 1962. One of Pawsey's last visitors was Fred Hoyle, who presented him with the prestigious Hughes Medal on behalf of the Royal Society of London (Lovell, 1964).

Pawsey's death marked the end of a golden era in Australian radio astronomy (Fig. 5.7). It also marked the end of the period of upheaval begun several years earlier, triggered by the decision to wind down support for the research programs of the 1950s and to direct the resources into the two giant instruments at Parkes and Culgoora. Several factors contributed to this turbulent period of transition. Paradoxically, one factor was the very success of the RP group. Although the size of the group had undergone a steady growth since the early post-war years, its key

Fig. 5.6 Joe Pawsey at the Vermilion River Observatory, Illinois, in September 1961, shortly before he was offered the position of Director of the US National Radio Astronomy Observatory (courtesy: CRAIA)

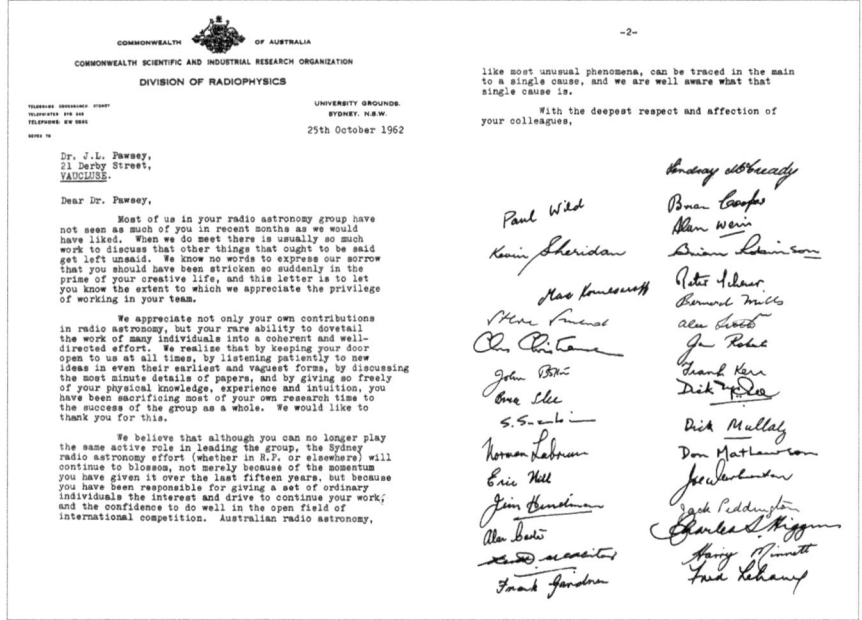

Fig. 5.7 A letter sent to Joe Pawsey, shortly before his death, by the radio astronomy group he had founded (courtesy: CRAIA)

personnel had changed remarkably little. With the group's growing international prestige and, until about 1960, its comfortable level of funding, there had been no incentive for its members to leave. The success of the group had given it a certain stability and cohesiveness. One measure of this success is that eight members of the group (Bolton, Christiansen, Mills, Minnett, Pawsey, Piddington, Robinson and Wild) were elected Fellows of the Australian Academy of Science, recognised as the highest honour Australia can bestow on its scientists. This success in many cases led to accelerated promotion through the ranks faster than most scientists in CSIRO and for some, at a relatively young age, to an eventual promotion barrier. The best prospects for advancement were then to be found beyond the doors of Radiophysics.

5.4   Pulsars – Cosmic Lighthouses at Radio Wavelengths

It is beyond the scope of this book to provide an overview of Australia's contributions to radio astronomy over the sixty-plus years since the Parkes Radio Telescope opened in 1961. However, the discovery and study of pulsars provides a good case study of how Australia has excelled in one particular branch of radio astronomy. Pulsars are also a good example of how Australia, with its clear view along the plane of the Milky Way, has had a significant advantage over the northern observatories. As former RP physicist Neville Fletcher once joked: "In astronomy circles it is often remarked – mostly by envious northerners – that God, in creating the Universe, perversely located the most interesting regions of our Galaxy in the southern hemisphere, but most of the astronomers in the north." (Robertson, 1992: 277).

In February 1968 the Cambridge group stunned the astronomical world by announcing it had discovered a completely new type of radio source. Inspired by the term quasar, the word 'pulsar' was coined for the new objects. Similar to other landmark discoveries in radio astronomy, the pulsar discovery was made by serendipity, not by design. Early in 1967 Antony Hewish and his graduate student Jocelyn Bell were observing quasars and other compact objects when Bell noticed a new type of radio source with fluctuations in its strength apparently unrelated to any type of local electrical interference. Further observations by Bell and Hewish established that the source lay in the plane of the Milky Way with signals consisting of a succession of regular pulses. The signal varied erratically in intensity, but the astonishing feature was the extraordinarily precise timing between individual pulses – exactly 1.3372795 seconds.

Within weeks three other sources had been discovered, each with a different pulsation period but each with the same incredible regularity. The Cambridge group thought it might have stumbled across signals sent by extra-terrestrial civilisations. The observations continued in secret for several months as Bell and Hewish worked to establish a natural, physical origin for the signals. Finally, convinced that the signals were not the creation of unknown alien life, Hewish announced the discovery (see Bell Burnell, 1977).

Fig. 5.8 In 1973 the Australian Government issued a new $50 note celebrating Australian science. One side featured the physiologist and Nobel Laureate Howard Florey and the other Ian Clunies Ross (above), the former Chairman of CSIRO (1949–59). The Parkes dish is at centre and above it is a contour map of the Large Magellanic Cloud and a solar burst recorded by the Radioheliograph at Culgoora. At far left is the signal from the pulsar CP 1919 recorded at Parkes in March 1968 (courtesy: Robertson collection)

Pulsar observations began at Parkes two weeks after the announcement of the discovery, principally by Max Komesaroff, Brian Robinson and visiting Indian astrophysicist Radhakrishnan. They began by correcting a small timing error in the pulse period of CP 1919, the first of the Cambridge pulsars (Robinson et al., 1968; see Fig. 5.8). In a study of the Vela pulsar (see below) they were able to measure a very small slowing down in its pulse period of 10 nanoseconds per day and also to show that each pulse was strongly polarised. Both these results provided evidence in favour of a 'lighthouse' theory put forward by Tom Gold at Cornell University. According to Gold, a pulsar consists of a neutron star spinning extremely rapidly and with its magnetic axis inclined at a large angle to the rotation axis (Gold, 1968; see Fig. 5.9). Another peculiar property of pulsars was also revealed by the Parkes observations on Vela. On occasions the pulse period of the pulsar would suddenly speed up. The word 'glitch' was coined to describe the abrupt change. After each glitch, the pulsar enters another long phase of gradual slowdown until the next glitch. The glitch is thought to be caused by a sudden change to the internal structure of the pulsar.

In the hunt for new pulsars Parkes was no match for the Molonglo SuperCross with its much larger collecting area at low frequencies. The Molonglo group lead by Michael Large discovered the first southern pulsars, including one which was shown to coincide with the position of the supernova remnant Vela X. A similar discovery by an American group of a pulsar at the centre of the famous Crab Nebula provided further evidence that some pulsars are created in supernova explosions of stars with masses much greater than the Sun. By the end of 1971 the Molonglo group, with a tally of 31, were the world leaders in pulsar discovery, while Jodrell Bank in

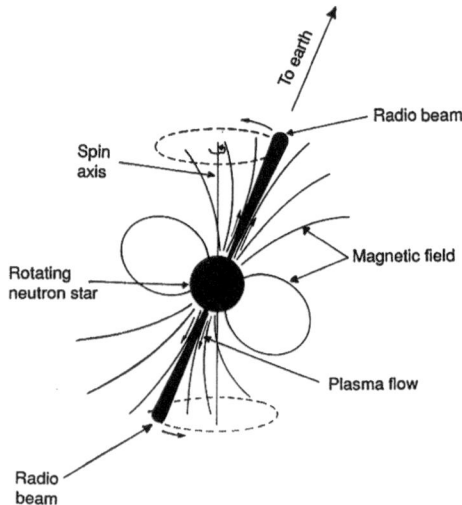

Fig. 5.9 A pulsar has its magnetic axis inclined to its spin axis. The intense magnetic field produces enormous forces that strip plasma material from the surface of the pulsar. The plasma is accelerated to nearly the speed of light causing it to emit intense radiation. Most of the radiation is beamed out along the north and south poles where the magnetic field is strongest. If the pulsar is suitably aligned in space, a pulse of radiation will be observed every time the beam sweeps across the Earth. The effect is the same as the flashing signal seen from a distant lighthouse (after Robertson, 1992: 311)

Cheshire, Green Bank in West Virginia and Arecibo in Puerto Rico were the most prominent groups in the northern hemisphere.

The Molonglo Cross was ideal for discovering pulsars, but not for studying them in detail. Observations of a particular pulsar required an instrument which could track the object for long periods of time. Here the Parkes dish came into its own in terms of actually learning what makes a pulsar tick. In 1976 Dick Manchester (Fig. 5.10) at Radiophysics led a collaboration between the Parkes and Molonglo groups in a survey of the southern sky. The great majority of pulsars were found to lie roughly along the Galactic Plane. In 1978 the collaboration published a catalogue of 155 pulsars, more than doubling the number of pulsars then known (Manchester et al., 1978).

In most branches of radio astronomy, advances in instrumentation usually lead to the next major breakthrough in the field. In 1997 a new multibeam receiver was fitted to the Parkes dish which allowed the simultaneous observation of 13 separate patches of sky and which lead to the discovery of over 1200 pulsars (Manchester et al., 2001). Among them was the discovery of what remains the only known 'double' pulsar – two pulsars bound in tight orbit around each other. The system radiates gravitational waves and is an ideal laboratory for carrying out precision tests of Einstein's general theory of relativity and other theories of gravity. In 2004 *Science* magazine voted the discovery as one of the top ten scientific breakthroughs of the year.

Fig. 5.10 Dick
Manchester is a world
authority on pulsars.
During his career he has
led teams that have
discovered about sixty per
cent of all known pulsars.
In 2019 Dick was awarded
the prestigious Matthew
Flinders Medal by the
Australian Academy of
Science (courtesy: CRAIA;
credit: Kristen Clarke)

Currently there are over 2700 known pulsars – with more than half discovered at Parkes – with pulse periods ranging from 1.4 milliseconds up to 23 seconds. Two major goals of pulsar researchers are to find the first sub-millisecond pulsar and to find a pulsar orbiting a black hole. The next chapter in the pulsar story is likely to be when the Square Kilometre Array (SKA) comes into operation in the late-2020s. With its exceptional sensitivity, it is expected that the SKA will be able to detect pulsars even in distant galaxies (see Section 5.7).

5.5   A Birthday Present for the Nation

In broad terms, the evolution of Australian radio astronomy can be divided into four phases or periods. The first period – the subject of this book – covered the years 1945–60 when the Radiophysics Lab not only dominated Australian radio astronomy, but became a global leader in the new field. It was characterised by small, relatively independent groups working at a variety of field stations in and around Sydney. In most cases these groups devised and built their own radio telescopes, all of which could be funded within the annual RP budget. As we saw earlier in this chapter, the second period of Australian radio astronomy came in the 1960s when Big Science arrived in Australia, with the Parkes dish, the Radioheliograph and the Molonglo SuperCross all funded by American philanthropy. For the first time the groups lead by Bernie Mills and Chris Christiansen at the University of Sydney challenged RP's leadership in the field. The third period in Australian radio astronomy began when 'aperture synthesis' became the dominant way of doing radio astronomy – the technique that uses the Earth's rotation and a number of radio antennas in an array to simulate or 'synthesise' the effect of observing with a very large single telescope. (We discuss the fourth period below.)

The first Australian synthesis telescope was brought into operation in the early 1970s at the Fleurs field station. As noted earlier, Chris Christiansen had managed to save his old 'Chris Cross' from the bulldozers by persuading Radiophysics to transfer ownership of the field station to the University of Sydney. The Chris Cross consisted of north–south and east–west arms, each comprising 32 dishes 5.8 m in diameter (see Orchiston and Mathewson, 2009). Christiansen added two larger (13.7 m) dishes to each arm to convert the grating cross into a very sensitive synthesis telescope with a 20 arcsec beam, the narrowest of any single radio telescope in the southern hemisphere. The Fleurs Synthesis Telescope (FST) was used during the 1970s to study the structure of large radio galaxies, supernova remnants and emission nebulae, and for a while it also continued to produce the daily 21 cm maps of the Sun. Although further upgrades were made in the 1980s, the retirement of Christiansen led to a winding back of the research program. The FST was shut down in 1988 on the very same day as the inauguration of the Australia Telescope (AT – see below).

The second Australian synthesis telescope was also developed at the University of Sydney. Similar to the first full-scale cross built at Fleurs in 1954, Bernie Mills knew that the SuperCross at Molonglo would only have a limited lifetime. As a transit instrument operating at the single frequency of 408 MHz, once the whole sky had been surveyed the scientific returns would diminish rapidly. The 'Molonglo Reference Catalogue' of over 12,000 strong sources and several catalogues of weaker sources was completed in 1978. (In comparison, the Parkes survey at 408 MHz carried out in the 1960s and followed by the survey at 2.7 GHz in the 1970s catalogued just over 8000 sources.) Mills then decided to transform the SuperCross from a transit telescope to a state-of-the-art synthesis telescope at the higher frequency of 843 MHz. The north–south arm was closed down and by 1981 the east–west arm, 1.6 km (one mile) in length, had been converted into a new steerable instrument, the Molonglo Observatory Synthesis Telescope (MOST), named after the nearby Molonglo River. The large collecting area and the long observing times ensured that, for most of the 1980s, MOST was the most sensitive radio telescope south of the equator. After several upgrades, MOST is still in operation today.

The success by Christiansen with the FST and by Mills with the MOST once again displayed their talent for developing innovative radio telescopes and techniques. It also called into question once again the wisdom of concentrating the resources at Radiophysics into a single multipurpose dish. However, by the mid-70s, after 15 years of operation, the Parkes Radio Telescope was undoubtedly an outstanding success. Each year increasing numbers of radio astronomers from Australian universities and from overseas competed for observing time. And yet there was a growing realisation that Parkes was falling behind the new generation of synthesis telescopes coming into operation in the northern hemisphere: the Westerbork Synthesis Radio Telescope built by the Dutch group had begun observations; the 5-km telescope built by Martin Ryle's group at Cambridge was operational; and the Very Large Array (VLA) was under construction in New Mexico by the National Radio Astronomy Observatory. With the development of computer technology, these instruments could produce images of radio sources of exceptionally high resolution, which were not possible with single dish instruments. At the same time, these arrays were able to operate at multiple frequencies and achieve high sensitivity through large collecting areas.

In 1975 a committee was formed to investigate the options for an Australian Synthesis Telescope (AST) (see Fig. 5.11), chaired by Paul Wild who had been appointed Chief of Radiophysics following the retirement of Taffy Bowen in 1971. The formation of the committee was an achievement in itself given the previous animosity between Radiophysics and the University of Sydney groups. Bolton wanted to add another large dish to the 64 m Parkes dish to form an instrument similar to the one he had designed at Owens Valley in California, while Christiansen was in favour of an array consisting of many small dishes. A compromise was eventually reached with the Parkes dish linked to a number of medium-sized dishes which could be moved along rail tracks (Fig. 5.12). There were, however, critics of the proposal who believed the AST design was only an incremental improvement on Australia's existing instruments. Others argued that a government funding cap of $A5.8 million for the AST was grossly inadequate for the project to succeed. With the astronomical community divided, the government decided not to fund the project.

Scientific proposals are often turned down because they are seen to be too ambitious or because they cost too much for the likely scientific returns. Oddly, the proposal for the AST failed because it was not ambitious enough and because it was hampered by the government's penny-pinching attitude. The turning point came in

Fig. 5.11 A meeting of the Australian Synthesis Steering Committee in 1977: (back) John Bolton (Parkes Observatory), Jon Ables (Radiophysics), Don Mathewson (Mt Stromlo and Siding Spring Observatories), Dick Manchester (RP), Bob Frater (U Sydney), Brian Robinson (RP), Paul Wild (chair, RP), Bernie Mills (U Sydney), Don Morton (Anglo-Australian Observatory), Ron Brown (Monash U) and Kel Wellington (RP). (front) Harry Minnett (RP), Robert Hanbury Brown (U Sydney), Chris Christiansen (U Sydney) and Kevin Westfold (Monash U) (courtesy: CRAIA)

Fig. 5.12 Sketch of the proposed Australian Synthesis Telescope (AST) which linked the Parkes dish with several medium-sized dishes mounted on rail tracks. Although the proposal was not funded, the idea evolved into plans for the Australia Telescope (courtesy: CRAIA)

Fig. 5.13 Bob Frater (left) was the driving force behind the Australia Telescope. In 1982 he was appointed Chief of the Radiophysics Lab, following David Martyn (1940), Fred White (1942), Taffy Bowen (1946), Paul Wild (1971) and Harry Minnett (1978). John Brooks (right) was appointed Project Manager during the construction of the AT and later became the Assistant Director of the Australia Telescope National Facility (courtesy: CRAIA)

1981 with the appointment of Robert (Bob) Frater as the new Radiophysics chief (Fig. 5.13). As the first chief recruited from outside CSIRO, Frater came from Christiansen's Electrical Engineering Department, where he had been part of the design teams for both the Molonglo SuperCross and the Fleurs Synthesis Telescope. Shortly before Frater's appointment, a mining company had announced plans for a new copper and gold mine less than 20 km away from the Parkes Observatory, raising the prospect of radio interference from its machinery seriously degrading the quality of the Parkes site (thankfully, the company later cancelled its plan). Following discussions between Frater and Brian Robinson, the head of the RP Radio Astronomy Group, the two realised that by moving any planned synthesis array away from Parkes would allow a much superior array of dishes to be built. The obvious place for a new site was Culgoora in northern New South Wales where Radiophysics operated the Radioheliograph and where much of the required infrastructure was already in place.

Over the following year the design of the new array evolved into one radically different to the AST. The main array at Culgoora would consist of five 22 m dishes mounted on a 3 km long east–west rail track and these dishes could be positioned at any one of 35 different stations along the track. A sixth dish, a further 3 km to the west, could also be moved along a short east–west track. A seventh stand-alone dish was positioned at Mopra Rock near Siding Spring Mountain, 130 km south from Culgoora, while a further 220 km south was the 64 m Parkes dish. Importantly, the three sites formed an almost perfect three-element array, since the separations between the sites were in the right ratio to produce excellent radio images.

The new design had a number of distinct advantages over the AST. First, it could accommodate the diverse objectives of the various Australian astronomy groups, such as those interested in wide-field mapping or very long baseline interferometry, and like the arrays in the northern hemisphere it could operate over a wide range of wavelengths (from 3 mm to 1 m). The single dish at Mopra Rock could also be used as a stand-alone instrument for millimetre wavelength observations of molecules in interstellar space, where previously radio astronomers had been reliant on the Parkes dish which could not operate below about 3 cm. The new design was in the same league as the new synthesis telescopes in the northern hemisphere and clearly capable of front-line research well into the twenty-first century. Not surprisingly, the new design won the strong support of Australia's astronomical community.

Funding the new telescope was another matter. Unlike the American philanthropic support for the Australian radio telescopes of the 1960s, the new array would be totally reliant on Federal Government funding. To help sell the project to Canberra, Bob Frater decided that at least 80% of the telescope would be built by Australian industry, and so stimulate the economy by promoting new and innovative technologies. A master stroke was the suggestion by David Jauncey to name the project the 'Australia Telescope', so that any serious opposition from politicians would seem almost unpatriotic.

An even better idea was to nominate the Australia Telescope (AT) to be a Bicentennial project in 1988. Canberra was organising the celebrations for the Bicentenary of European settlement in 1788 and wanted to support high profile projects to mark the occasion. The appeal of the AT was that not only would it consolidate Australia's position as a world leader in astronomy, but would also be a fine tribute to its astronomical heritage. After all, Captain James Cook and his crew were sent to the South Pacific in 1768 to observe the transit of Venus (see our Frontispiece on page v and Orchiston, 2017) and then to chart the east coast of the fabled southern land, not to mention Aboriginal astronomy stretching back tens of thousands of

years (for a primer see Norris and Norris, 2009). In August 1982 the Federal
Government approved funding of up to $25 million for the construction of the AT,
over four times the earlier miserly allocation for the AST. Fortunately, the funds
were indexed to inflation which ran high throughout the 1980s. The final cost of the
AT was just over $50 million in 1988 dollars (for detailed accounts of the planning
and construction see Frater, 2019; Haynes et al., 1996; see also Sim, 2013).

To free up operational funding for the AT, Frater made the difficult decision in
1982 to close down the Radioheliograph at Culgoora. It was a controversial decision, but even its founder Paul Wild agreed that after 15 years of successful operation the instrument was coming to the end of its scientific life (the Culgoora site was
later renamed the Paul Wild Observatory). Site preparation work for the AT began
in September 1984 and rapid progress was made in laying the wide-gauge (9.6 m)
rail track to support the six movable dishes (Fig. 5.14). Despite careful planning,
there were some unforeseen environmental problems associated with this rural site,
ranging from kangaroos bounding over the metre-high perimeter fence to sulphur-
crested cockatoos chewing holes through the fibreglass insulation and aluminium
foil used on each of the dishes.

By December 1985 all civil works at Culgoora were complete and work began
on the construction of the 22 m dishes. The inner 15 m surface of each dish was
solid aluminium with the outer section perforated aluminium to reduce weight and

Fig. 5.14 The Australia Telescope Compact Array (ATCA) under construction at Culgoora in
1987. The extraordinary surface accuracy of each 22 m dish enables the array to operate at wavelengths as short as 3 mm (courtesy: CRAIA)

wind loading. To operate at wavelengths as short as 3 mm (100 GHz) required surface accuracies of 0.15 mm for the inner panels and 0.25 mm for the exterior ones. To get this level of precision RP staff developed an innovative technique for forming the parabolic surface that was over ten times cheaper to manufacture than the conventional and expensive method of using machined moulds. Later, Radiophysics and Macdonald Wagner, the consulting engineers building the AT, formed a consortium to commercialise the RP technique and it was used in a variety of satellite and telecommunications dishes elsewhere in Australia and in countries such as Cambodia and Vietnam. The total value of these commercial contracts was estimated to exceed the total cost of the Australia Telescope. In all, over 30 companies were involved in the construction and the project was able to comfortably meet Bob Frater's pledge of at least 80% Australian content.

   In the lead up to the inauguration, a major organisational change came with the formation of the Australia Telescope National Facility (ATNF) which would manage the Australia Telescope operations, including the Parkes Observatory. The ATNF was to have essentially the same status as other CSIRO Divisions and, in effect, it meant that radio astronomy would be removed from Radiophysics where it had been the major research program since the late 1940s. In practice, there was relatively little impact on the Radio Astronomy Group itself, as Radiophysics and the ATNF were to be co-located on the same site in the suburb of Marsfield in north–west Sydney (in 1968 the Radiophysics Lab had moved from its original building in the grounds of the University of Sydney to this six-hectare greenfield site). After a worldwide search Professor Ron Ekers (Fig. 5.15) was appointed the Foundation Director of the Australia Telescope. After completing his PhD in 1967 at the Australian National University (co-supervised by John Bolton), Ekers held positions at Caltech, Cambridge and the Kapteyn Observatory in the Netherlands. In 1980 Ron was appointed the inaugural Director of the Very Large Array in New Mexico, built and operated by the National Radio Astronomy Observatory. Ekers brought a wealth of experience in image processing and had been a leading advocate of building a major synthesis telescope in the Southern Hemisphere.

Fig. 5.15 Professor Ron Ekers (AO) was the Foundation Director (1988–2003) of the Australia Telescope National Facility. Ron is the only Australian-born astronomer to be elected President of the International Astronomical Union (2003–2006) (courtesy: CRAIA)

Fig. 5.16 (above) Five of the six 22 m dishes making up the ATCA. The sixth dish is 3 km to the west. (below) The Australia Telescope was officially opened by Prime Minister Bob Hawke in September 1988, as part of Australia's Bicentennial celebrations. At left are Hazel Hawke and science minister Barry Jones (courtesy: CRAIA)

The official opening ceremony of the Australia Telescope was held on a very windy day in September 1988. Although the main construction was complete, the project was in fact about 12 months behind the original schedule with much work remaining to get the telescope fully operational. Grote Reber and a Who's Who of

Australian astronomy were among the 800 guests. The Prime Minister Bob Hawke and wife Hazel were 'driven' to the official dais on one of the 22 m dishes (Fig. 5.16). After the speeches, Hawke declared the Australia Telescope open and then pressed a button causing three of the dishes to tilt down towards the guests and release a swarm of green and gold balloons.

*****

A brief interlude. As we saw in Chapter 4, the Radiophysics group of John Bolton, Gordon Stanley and Bruce Slee at the Dover Heights field station were able to show that the intense radiation from the northern Cygnus constellation came from a compact, point-like source. In June 1947 the group thought that they had found a second source in the southern constellation of Centaurus, but repeated attempts to confirm the detection proved a frustrating failure. The sensitivity of their equipment was simply not good enough to detect sources any fainter than the strong Cygnus source. Gordon Stanley made the crucial breakthrough in October 1947 when he was able to eliminate most of the noise variations in the receivers, so that much fainter signals could be detected. Early in November, the Dover Heights team was rewarded by the detection of a second source in the Taurus constellation, followed by a third in Virgo and then – the one that had eluded them – a fourth in Centaurus. The striking images of Cygnus A and Centaurus A in Fig. 5.17 are a powerful illustration of just how far radio astronomy has come in such a short period of time.

5.6   A Radio Telescope as Wide as Australia

The technique of interferometry has been an integral part of Australian radio astronomy from the very beginning. As we saw in Chapter 2, in 1946 Joe Pawsey, Lindsay McCready and Ruby Payne-Scott used a sea interferometer at Collaroy to show that sunspots are an intense source of radio emission. Similarly, in 1949 Bernie Mills at the Badgery's Creek field station constructed an interferometer consisting of three broadside antennas to study the strongest discrete radio sources (Fig. 2.53). The antennas were connected to each other using underground cables and their signals were sent to a receiver hut where they were combined to produce the distinctive interference patterns. Later, in a world first, Mills used a microwave link to connect one of the broadside antennas to another movable antenna mounted on a trailer (Fig. 2.55). The distance between the two antennas, known as the baseline, could be varied by up to 10 km.

A major breakthrough in interferometry took place in the late 1960s in Canada and the US with the introduction of atomic clocks as time and frequency standards, with a precision better than one part in 10^{12}. Rather than rely on cables or microwave links between the antennas in an interferometer, atomic clocks could be used to precisely synchronise the signals received at each antenna. The received signals could be recorded on magnetic tape and then the tapes from each antenna later combined ('correlated') at a central processing station. The interference patterns were

Fig. 5.17 Two old favourites in a new light: (above) A multi-wavelength view of the spectacular active galaxy Cygnus A, featuring X-ray data (blue), radio emission (red) and optical observations

Fig. 5.18 David Jauncey at the ATNF pioneered very long baseline interferometry in Australia. He used the 64 m dish (later upgraded to 70 m and shown at centre) at the NASA tracking station at Tidbinbilla, south–west of Canberra, combined with other radio telescopes in Australia and the United States (courtesy: CRAIA)

no longer produced in real time, and in principle there was no longer a limit to the possible distances between the antennas. This marked the start of a major new branch of radio astronomy – very long baseline interferometry (VLBI).

The first step in Australia towards VLBI was in 1967 by a group at the Defence Science and Technology Organisation (DSTO) in Adelaide. The group used the 26 m NASA tracking dishes at Island Lagoon near Woomera and at Tidbinbilla near Canberra to study the internal structure of the quasar 3C273. The group also carried out a successful trans-Pacific experiment connecting a NASA dish in California with the Tidbinbilla dish, but in 1973 DSTO decided to discontinue this pioneering work in VLBI. The next step towards Australian VLBI was the appointment of David Jauncey to the Radiophysics Lab in 1974 (Fig. 5.18). After graduating from the University of Sydney, Jauncey had spent almost ten years at Cornell University where he joined a group carrying out the first VLBI experiments in the United

---

◄

Fig. 5.17 (continued) by the Hubble Telescope (white). In the early 1950s, the two radio jets emanating from the central core were mistakenly thought to be two galaxies in collision (see Fig. 4.15) (courtesy: NASA and seven other US organisations). (below) A composite image showing the radio glow from the galaxy Centaurus A in comparison with the full Moon. The observations were made at 1.4 GHz by the AT Compact Array and the Parkes Radio Telescope. The huge extent of the glow dwarfs the size of its optical counterpart NGC 5128. The white dots in the sky are not stars, but represent radio sources such as quasars and radio galaxies in the distant Universe [courtesy: Ilana Feain, Tim Cornwell, Ron Ekers, Shaun Amy (CSIRO); R. Morganti (ASTRON); N. Junkes (MPIfR)]

States. On his return Jauncey moved to Canberra, rather than Sydney, to take advantage of a new 64 m antenna NASA had constructed at Tidbinbilla as part of its Deep Space Network (DSN) for tracking spacecraft. Under Australia's agreement with NASA, a small fraction of the antenna's time would be available to Australian astronomers. Jauncey began a collaboration with a group led by Robert Preston at the Jet Propulsion Lab in Pasadena that used Tidbinbilla and other NASA facilities worldwide to study compact, milliarcsecond components in quasars and radio galaxies. The whole sky survey ran for almost ten years and lead to a catalogue of over 900 compact radio sources with positions measured to sub-arcsecond precision (Jauncey, 2012).

By about 1980 it became clear to the growing band of radio astronomers involved in VLBI that Australia needed its own VLBI imaging capability. This led to the formation the Southern HEmisphere VLBI Experiment (SHEVE) consisting of five antennas at Parkes, Fleurs, Tidbinbilla, Hobart and Alice Springs, with a sixth at Hartebeesthoek in South Africa. SHEVE was a cooperative effort involving both astronomers and scientists involved in geodesy – the study of the precise size and shape of the Earth. The time synchronisation was done with a rubidium atomic clock supplied by Geoscience Australia, the recording equipment was on loan from the Jet Propulsion Lab, and the correlation of the data tapes was carried out at a joint Caltech–JPL facility in Pasadena. As an indication of the increasingly international and cooperative nature of radio astronomy, especially VLBI, the first series of publications by the SHEVE collaboration, reporting observations made in April 1982, numbered 29 authors from 14 institutions (see e.g. Jauncey et al., 1983).

The next major step in Australian VLBI came in 1984 with the proposal by Ray Norris (Fig. 5.19) to use the existing radio link between the 64 m dishes at Parkes and Tidbinbilla to form a new interferometer. Before the 64 m Tidbinbilla dish came into operation in 1973, NASA relied on 26 m dishes at Tidbinbilla and at Honeysuckle Creek (also near Canberra) to communicate with the Apollo 11 spacecraft in 1969

Fig 5.19 Three of the main players in the development of Australian VLBI (from left): Ray Norris, John Reynolds and Tasso Tzioumis (courtesy: CRAIA; credit: John Sarkissian)

through to Apollo 17 in 1972. NASA contracted the Parkes dish during each Apollo mission in order to use its much larger collecting area to receive the TV signals from the lunar surface (in fact, the world saw the first moon walk by Neil Armstrong and Buzz Aldrin from signals received by both Honeysuckle Creek and Parkes). Norris and Mike Kesteven were able to use the Apollo radio link (and not atomic clocks), along with new electronics and software, to form the Parkes–Tidbinbilla Interferometer (PTI). With a baseline of 275 km the PTI became the world's longest baseline interferometer operating in real time and was especially well-suited for studying rapidly varying and transient radio sources (Norris and Kesteven, 2013).

Australian interferometers were involved in two of the most exciting discoveries of the 1980s. During the 1960s several Parkes radio sources held the record for the most distant known object in the Universe, including the quasar PKS 0237–23 with a red-shift of 2.22. By 1973 the record had passed to two sources observed by John Kraus and his group at Ohio State University, with redshifts of first 3.40 and then 3.53. Almost a decade passed without the record being broken and many astronomers thought this indicated that the Ohio sources marked the edge of the observable Universe.

The identification and redshift measurement of PKS 2000–330 in March 1982 by Bruce Peterson, Ann Savage, David Jauncey and Alan Wright involved a series of observations with four telescopes, two radio and two optical. The source itself had been detected at Parkes as early as 1970, but there had been nothing to suggest anything unusual. In 1978 the radio position of the source was measured to greater accuracy with the Tidbinbilla short baseline interferometer (consisting of the 64 m and 34 m NASA dishes). Next it was possible to reliably identify the source on a photographic plate taken with the UK Schmidt telescope at Siding Spring in NSW. Finally, observations with the Anglo-Australian telescope at Siding Spring revealed a new record redshift of 3.78 for PKS 2000–330, making it not only the most distant but also the most luminous object known in the Universe (Peterson et al., 1982). The discovery received widespread media coverage internationally, including front-page headlines in several Australian newspapers.

Possibly the most important astronomical event of the 1980s was the explosion of the supernova SN1987A in February 1987. The explosion occurred in the Large Magellanic Cloud and despite the distance it became the first naked-eye supernova since Kepler's star in the year 1604. Immediately after the discovery, virtually every telescope in the Southern Hemisphere – radio, optical and infrared – focussed on the LMC. Duncan Campbell-Wilson at the Molonglo Cross was the first to detect radio emission from the supernova, at 843 MHz, two days after its discovery. Observations with high angular resolution were essential to separate the supernova emission from the crowded radio background, an ideal challenge for the Parkes–Tidbinbilla Interferometer, which had just come into operation. PTI observations at 2.29 and 8.41 GHz were underway almost immediately and four days later the radio emission peaked but then, a few days later, it faded and became undetectable (Turtle et al., 1987). Although the supernova remnant has been monitored regularly over the years (see Fig. 5.20), and most recently by the Parkes dish (Zhang et al., 2018), there is still no evidence for a pulsar. It is quite possible that the explosion created a pulsar, but with an orientation where the radio beam does not sweep across the Earth.

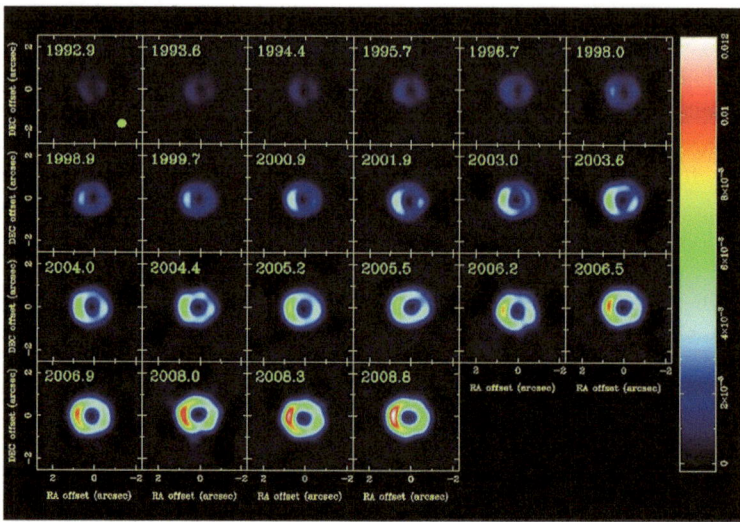

Fig. 5.20 The evolution of the supernova remnant 1987A observed by the AT Compact Array at 9 and 18 GHz from 1992 to 2008. The remnant has grown in size and brightened significantly as the shock wave from the supernova explosion collides with material ejected in the death throes of the star (after Ng et al., 2008)

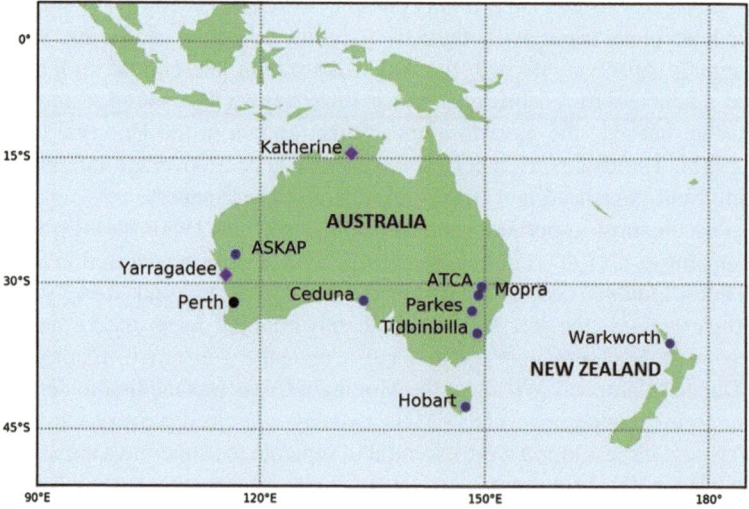

Fig. 5.21 The Australian Long Baseline Array currently consists of eleven antennas, including one at Hartebeesthoek in South Africa. Yarragadee and Katherine are part of the AuScope geodetic VLBI array, operated in partnership with Geoscience Australia. The Australia Telescope National Facility operates ATCA, Mopra, Parkes and the Australian Square Kilometre Array Pathfinder (ASKAP) in Western Australia (see next section). Tidbinbilla is operated by CSIRO on behalf of NASA. Hobart and Ceduna are operated by the University of Tasmania and Warkworth by the Auckland University of Technology. The correlation of the signals from each antenna is carried out by CSIRO at the Pawsey Supercomputing Centre in Perth, Western Australia (courtesy: Chris Phillips, ATNF)

The current VLBI facility in Australia is known as the Long Baseline Array (Fig. 5.21). Unlike the SHEVE collaboration in the 1980s, where most of the recording equipment was borrowed from the Jet Propulsion Lab, the LBA is fully funded by the Federal Government's Major National Research Facilities program. The LBA has the status of a national facility, managed by the Australia Telescope National Facility, and so potential users are granted observing time according to the scientific merit of their proposal. The LBA is also an important element in the Asia–Pacific region, linking Australia to radio telescopes in Brazil, Chile, China, India, Japan, New Zealand and South Korea. Finally, the LBA was also involved in what is known as space VLBI (SVLBI) where ground-based arrays are linked to antennas in Earth orbit to create baselines as long as several Earth diameters. Two of the main Australian SVLBI collaborations have been the Russian-led RadioAstron and the Japanese-led VLBI Space Observatory Program, involving up to fifty radio telescopes in more than twenty countries. VLBI and SVLBI have been spectacular examples of how progress in science is achieved through international collaboration.

5.7   Simply Astronomical – the Square Kilometre Array

Until recently, with the exception of the small group at the University of Tasmania (Fig. 5.22), Australian radio astronomy has really been New South Wales radio astronomy. Every major Australian radio telescope was located in NSW, lying roughly on a north–south line running down the centre of the state, west of the Great Dividing Range of mountains. Towards the north near Narrabri were first the Radioheliograph followed by the Australia Telescope Compact Array, and then to the south was the small precision dish near Siding Spring. Further south was the Parkes dish and then another four-hour drive to the Molonglo Cross near Canberra. The one exception to the north–south line was the Fleurs Synthesis Telescope near Sydney, operated by Chris Christiansen and his group. Australia's optical astronomers were also concentrated in NSW. The intensity interferometer built by Robert Hanbury Brown, and used for measuring the diameter of stars, was located at Narrabri, while Siding Spring mountain was host to a suite of instruments operated by the Australian National University, as well as the UK 1.2 m Schmidt Telescope (opened in 1973) and the Anglo-Australian 3.9 m Telescope (opened in 1974).

Early in the new millennium the focus of Australian radio astronomy began to shift from NSW to a remote site in outback Western Australia. The site in Murchison Shire north of Perth had been shortlisted as a possible host for the largest, the most complex, and by far the most costly radio telescope ever planned – the Square Kilometre Array.

But first a brief diversion. Much of this chapter has been devoted to the evolution of Australian radio telescopes over the past fifty or so years and very little has been said about the more general development of Australian astronomy as a profession. In the 1950s there was no national association of astronomers and so if any of the

Fig. 5.22 The Physics Department at the University of Tasmania currently maintains two radio observatories, one at Mt Pleasant near Hobart (shown here), and the other at Ceduna on the Nullabor Plain (Fig. 5.21). At both stations, former US satellite-tracking 23 m dishes are the main antennas (courtesy: University of Tasmania and Martin George)

radio astronomers at Radiophysics did feel part of a particular professional discipline it was more likely to be physics or radio engineering. This changed in 1966 when a small group of astronomers in Sydney formed the Astronomical Society of Australia (ASA), with Harley Wood, Director of Sydney Observatory, elected the inaugural president (see Fig. 1.17). The initial membership was about 90 and the great majority were either radio astronomers from Radiophysics and the University of Sydney or optical astronomers from the Mt Stromlo Observatory (a department of the Australian National University) (Lomb, 2015).

The membership of the ASA grew steadily to about 350 by the year 2000 and since then it has doubled to over 700 in 2019. This growth is impressive when compared with other Australian professional bodies in the physical sciences which have only shown modest growth, or indeed flat-lined, over the same period of time. Also impressive is that the number of women ASA members has grown from a handful in 1966 to 30% of the current membership (Wyithe, 2019). This expansion also reflects the growth and diversification of Australian astronomy with significant new groups at universities and institutions across the country: the Australian Astronomical

Observatory (formerly the Anglo-Australian Observatory), University of NSW, Macquarie University, Western Sydney University (NSW); University of Queensland, University of Southern Queensland (QLD); University of Adelaide (SA); Monash University, Swinburne University, University of Melbourne (VIC); and Curtin University, International Centre for Radio Astronomy Research, University of Western Australia (WA).

The Square Kilometre Array (SKA) is the astronomical equivalent of the Large Hadron Collider, the particle accelerator operated by the European Organisation for Nuclear Research (CERN) in Switzerland, the most expensive scientific instrument on the planet. The SKA will be too complex and too costly to be built by any one country. Instead, similar to CERN which is a consortium of mainly European countries, the SKA will be built by an international consortium of countries, consisting of (at the time of writing) Australia and fifteen other countries. The origins of the SKA go back to about 1990 when radio astronomers debated what type of radio telescope would be needed to investigate the leading astronomical questions of the new millennium. The consensus was that a new telescope would need to have 50 times the sensitivity of any existing telescope and with a total collecting area of one square kilometre (Ekers, 2013). An international working group was established in 1993, followed by the International SKA Project Office in 2003. Next came the SKA Program Development Office, which oversaw much of the development work up to the establishment of the SKA project as a legal entity in 2011 (Fig. 5.23). Four bids to host the SKA were short-listed – Australia, China, South Africa and a joint Argentina–Brazil bid – and in 2008 the list was whittled back to just Australia and South Africa.

Most commentators thought Australia's bid to host the SKA was much stronger than South Africa's. South Africa planned to locate most of the SKA in the semi-arid Karoo region of the country, but to achieve long baselines parts of the giant array would need to be located in eight other countries in southern Africa, including

Fig. 5.23   Richard Schilizzi was the inaugural Director (2003–2011) of the International SKA Project Office, based initially in the Netherlands and from 2008 at the University of Manchester. Richard completed a PhD on the structure of extragalactic radio sources at the University of Sydney in 1973 (courtesy: Richard Schilizzi)

Botswana, Namibia and Zambia. In contrast, Australia was large enough geographically to accommodate all of the SKA (with the possible inclusion of New Zealand) and so the logistics of establishing and operating the telescope would be much easier. Australia offered political stability with much of the infrastructure in place and, not least, was already firmly established as a leader in radio astronomy. It therefore came as a great surprise in May 2012 when it was announced that the SKA would not be awarded to just one country, but be shared between Australia and the consortium of southern African countries. Rather than one country or region, there would be two. Australia would host an array operating at low frequencies (50 to 350 MHz) – known for short as SKA-low – while southern Africa would host a completely different array operating at frequencies from 350 MHz to 14 GHz – known as SKA-mid.

The site chosen to host SKA-low is a flat semi-arid plain in Murchison Shire, approximately 800 kilometres north of Perth in central Western Australia. In 2009 CSIRO bought the pastoral lease to Boolardy Station in the middle of the shire, relocated the cattle and fenced off the site. Although the shire itself is about the size of the Netherlands, there are only about 110 permanent residents most of whom belong to the Wajarri Yamatji nation. The site has exceptionally low levels of manmade radio interference and to preserve the quietness a 520 kilometre exclusion zone has been established that bans any possible sources of radio interference (apart from satellite emissions – see Fig. 5.24).

Fig. 5.24 The Project Scientist for the ASKAP, Lisa Harvey-Smith, at the entrance to the observatory. A visit to the Murchison Radio-astronomy Observatory (MRO) from Sydney involves two full days of travel. First is a four-hour flight to Perth, followed by a one-hour flight north to the coastal town of Geraldton. Next day a 4WD vehicle, loaded with supplies, is required to drive north–east along a 300 km stretch of dirt road to the observatory site (courtesy: Lisa Harvey-Smith)

In the lead up to the construction of the SKA, three so-called 'precursor' arrays have been built, one in South Africa and two at the Murchison Radio-astronomy Observatory (MRO). These arrays are exceptionally powerful telescopes in their own right, but have the dual role of being test-beds to perfect the technology and techniques required by the SKA. MeerKAT at the Karoo site in South Africa consists of sixty-four 13.5 m dishes, covering the frequency range 580 MHz to 14 GHz, and came into operation in 2018. The Australian Square Kilometre Array Pathfinder (ASKAP) consists of thirty-six 12 m dishes and includes innovative technologies such as phased array feeds to give a wide field of view (30 square degrees). ASKAP was designed and built by CSIRO and officially opened in October 2012 (though it took several years for ASKAP to become fully operational – see Figs 5.25 and 5.26).

Fig. 5.25 Some of the thirty-six dishes forming the Australian Square Kilometre Array Pathfinder. The full array is spread over an area 6 km wide. ASKAP was designed and built by CSIRO at a cost of $A188 million (courtesy: CRAIA and CSIRO/Alex Cherny; all rights reserved)

Fig. 5.26 The Murchison Radio-astronomy Observatory and ASKAP were officially opened in October 2012. A highlight were dances performed by the traditional owners of the land (courtesy: CRAIA; credit: Carole Jackson)

A wide range of observations have now been completed, mainly high-speed large-scale surveys of the southern sky, where it seems almost obligatory to name the project so that its acronym corresponds to an Australian animal: Evolutionary Map of the Universe (EMU), Deep Investigations of Neutral Gas Origins (DINGO), and we'll let our readers puzzle over what the projects POSSUM and WALLABY could possibly be.

The second Australian SKA precursor is the Murchison Widefield Array (MWA), a low frequency (80–300 MHz) interferometer which came into operation in 2013 (Fig. 5.27). In contrast to ASKAP, the MWA project is an international collaboration currently consisting of 21 institutions in six countries (Australia, Canada, China, Japan, the United Arab Emirates and the United States), with funding for the MWA provided by the partner institutions and a number of national funding agencies. Eight of the 21 partner institutions are Australian with the leading roles played by Curtin University and the International Centre for Radio Astronomy Research in Perth. ICRAR was formed in 2009 with the specific purpose of supporting Australia's bid to host the SKA and currently consists of over 100 staff and post-graduate students (Fig. 5.28). The historical concentration of Australian radio

Fig. 5.27 The Murchison Widefield Array consists of thousands of 'bowtie' dipoles arranged in groups of sixteen, known as tiles. The tiles are scattered throughout the observatory. The MWA has carried out a full survey of the southern sky over the frequency range 72–231 MHz, producing a catalogue of 300,000 radio sources (courtesy: ICRAR/Curtin University)

astronomy in Sydney at Radiophysics, the University of Sydney, and then the Australia Telescope National Facility, has now shifted to a quite significant degree to ICRAR and Curtin University in Perth.

The construction of SKA-Low – or the Low Frequency Aperture Array (LFAA) as it is now known – is expected to begin in early 2022. Phase 1 of the construction will see the deployment of over 130,000 identical antennas to create what has been

Fig. 5.28 Rachel Webster (left) at the University of Melbourne initiated and guided plans for the Murchison Widefield Array. The current MWA Director Steven Tingay is based at the International Centre for Radio Astronomy Research in Perth (courtesy: Rachel Webster and ICRAR; credit: MCB Photographics)

described as 'a forest of steel Christmas trees' (Mannix, 2018) (see Figs 5.29 and 5.30). The core of the array will be tightly packed with about three-quarters of the antennas located within a 2 km radius. The remaining antennas will form spiral arms spanning about 50 km to further enhance the image detail. The torrent of data produced by LFAA will be sent by fibre-optic cables to Geraldton on the coast and then south to the Pawsey Supercomputer Centre in Perth, named in honour of the founding father of Australian radio astronomy. Processing and storing the extraordinary amounts of data generated will require colossal computer power – at peak operation it is estimated that the LFAA could generate more data in one day than the entire current global internet.

In the meantime, at the SKA headquarters at Jodrell Bank in England (Fig. 5.31) attempts are being made to finalise the membership of the SKA Organisation, which coordinates about 100 organisations across twenty countries involved in the design and development of the SKA. Currently there are the three host countries Australia, South Africa and the UK, and another thirteen full-member countries Canada, China, France, Germany, India, Italy, Japan, the Netherlands, Portugal, South Korea, Spain, Sweden and Switzerland, with further full-member countries expected to join in the future. One setback has been the announcement by New Zealand in July 2019 that it was quitting the SKA, arguing that the cost of its membership would not directly benefit enough of its radio astronomers (Cartlidge, 2018). Originally, there were plans for both Australia and New Zealand to host the SKA, but this fell through in 2012 with the decision to divide the project between Australia and South Africa. However, with at least eight other countries expressing an interest in joining the SKA Organisation, there may be enough funding to cover the projected price tag of about two billion euros, so that the first phase of the SKA can be completed by the end of the 2020s.

*****

Earlier in this chapter we noted that, in broad terms, the evolution of Australian radio astronomy can be divided into four phases or periods. The first period covered the years 1945–60 when the Radiophysics Lab emerged as a global leader in the new field. It was characterised by small, relatively independent groups working at a variety of field stations in and around Sydney. In most cases these groups devised and built their own radio

Fig. 5.29 (above) The Aperture Array Verification System (AAVS) is a testbed for optimising the design of the SKA-Low antennas. (below) SKA-Low engineer Maria Grazia Labate inspects one of the 256 'Christmas tree' antennas making up the AAVS. The antennas have been designed to mini-mise cost and to maximise reliability in their harsh climate (courtesy: ICRAR/Curtin University)

telescopes, all of which could be funded within the annual RP budget. The second period of Australian radio astronomy came in the 1960s when Big Science arrived in Australia, with the Parkes dish, the Radioheliograph and the Molonglo Cross all funded by American philanthropy. For the first time the groups lead by Bernie Mills and Chris Christiansen at the University of Sydney challenged RP's dominance in the field. The third period began in the 1980s when the Australia Telescope bicentennial project revit-alised Australian radio astronomy. The development of Australian very long baseline interferometry in the 1980s also belongs to this third period.

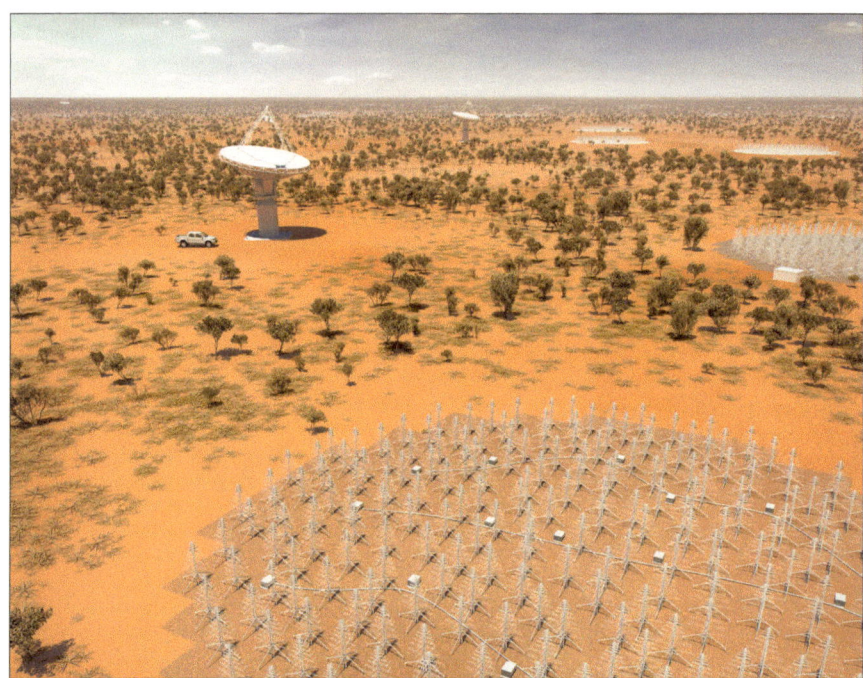

Fig. 5.30 Computer image showing how the LFAA might appear after the full deployment of the 130,000 tree antennas in Phase 1. Eventually, provided Phase 2 is fully funded, the number of antennas will be increased to about one million during the 2030s (courtesy: SKA Observatory)

Fig. 5.31 The SKA Organisation is housed in a new building at Jodrell Bank. The organisation planning the world's largest radio telescope is next door to the very first giant radio telescope, completed in 1957. In 1987 the 250 ft dish was renamed the Lovell Telescope in honour of Sir Bernard. In February 2021, the SKA Organisation was superseded by the SKA Observatory, with Philip Diamond as the inaugural Director-General. The SKA Observatory is the intergovernmental organisation that will build and operate the SKA telescopes (courtesy: SKA Observatory)

We are now very much in the fourth period where Australia will play host to a gigantic multinational array, where both the cost and the sensitivity of the SKA will be up to two orders of magnitude greater than the Australia Telescope Compact Array. Will there ever be a fifth period of Australian radio astronomy? With the SKA projected to have a lifespan of fifty years or more, it is difficult to say. If there is to be a fifth period, then almost certainly it will be the result of major advances in the technology of radio telescopes.

Over sixty years after its inauguration in 1961, the Parkes Observatory holds the current Australian record for longevity. Perhaps one day that record will pass to the Murchison Radio-astronomy Observatory.

References

Bell Burnell, S.J., 1977. Little Green Men, white dwarfs, or pulsars? *Annals of the New York Academy of Science*, 302, 685–689.

Bracewell, R.N., 2005. Radio astronomy at Stanford. *Journal of Astronomical History and Heritage*, 8, 75–86.

Campbell, D.B., 2019. Radio astronomy at Cornell University: the early years, 1946 to 1962. *Journal of Astronomical History and Heritage*, 22, 503–520.

Cartlidge, E., 2018. New Zealand pulls out of the Square Kilometre Array. *Physics World*, August, p. 8.

Christiansen, W.N., 1990. Letter to Peter Robertson, dated 27 February.

Cohen, M.H., 1994. The Owens Valley Radio Observatory: The early years. *Engineering and Science* 57(3), 9–23.

Edge, D.O., and Mulkay, M.J., 1976. *Astronomy Transformed: The Emergence of Radio Astronomy in Britain*. New York, Wiley.

Ekers, R., 2013. The history of the Square Kilometre Array (SKA) born global. Proceedings of the Conference on Resolving the Sky – Radio Interferometry: Past, Present and Future. Manchester, UK, April 2012.

Frater, R.H., 2019. Chapter 5: CSIRO 1981–1999. A memoir, in preparation.

Frater, R.H., Goss, W.M., and Wendt, H.W., 2017. *Four Pillars of Radio Astronomy: Mills, Christiansen, Wild, Bracewell*, Cham, Switzerland, Springer.

George, M., Orchiston, W., Slee, B., and Wielebinski, R., 2015. The history of early low frequency radio astronomy in Australia. 2: Tasmania. *Journal of Astronomical History and Heritage*, 18, 14–22.

George, M., Orchiston, W., and Wielebinski, R., 2017. The history of early low frequency radio astronomy in Australia. 8: Grote Reber and the 'Square Kilometre Array' near Bothwell, in the 1960s and 1970s. *Journal of Astronomical History and Heritage*, 20, 195–210.

Gold, T., 1968. Rotating neutron stars as the origin of the pulsating radio sources. *Nature*, 218, 731–732.

Hanbury Brown, R., 1960. Letter to J.L. Pawsey, dated 7 June. National Archives of Australia, file Z3.

Haynes, R., Haynes, R., Malin, D., and McGee, R., 1996. *Explorers of the Southern Sky: A History of Australian Astronomy*. Cambridge, Cambridge University Press.

Jauncey, D.L., 2012. SHEVE '82, Tasso, and the coming age of Southern Hemisphere VLBI. Proceedings of the Conference on From Antikythera to the Square Kilometre Array: Lessons from the Ancients. Kerastari, Greece.

Jauncey, D.L., Preston, R.A., and (23 others), 1983. SHEVE: The Southern Hemisphere VLBI experiment. *Bulletin of the American Astronomical Society*, 15, 648.

Kellermann, K.I., Bouton, E.N., and Brandt, S.S., 2020. *Open Skies: The National Radio Astronomy Observatory and Its Impact on US Radio Astronomy*. Cham (Switzerland), Springer.

Kellermann, K.I., Orchiston, W., and Slee, B., 2005. Gordon James Stanley and the early development of radio astronomy in Australia and the United States. *Publications of the Astronomical
Society of Australia*, 22, 13–23.

Lomb, N., 2015. How astronomers focussed the scope of their discussions: The formation of the
Astronomical Society of Australia. *Historical Records of Australian Science*, 26, 36–57.

Lovell, B., 1964. Memoir of J.L. Pawsey. *Biographical Memoirs of the Royal Society of London*,
10, 229–43.

Lovell, B., 1968. *The Story of Jodrell Bank*. New York, Harper & Row.

Manchester, R.N., Lyne, A.G., Taylor, J.H., Durdin, J.M., Large, M.I., and Little, A.G., 1978.
The second Molonglo pulsar survey – discovery of 155 pulsars. *Monthly Notices of the Royal
Astronomical Society*, 185, 409–421.

Manchester, R.N., Lyne, A.G., Camilo, F., and 9 others, 2001. The Parkes multibeam pulsar survey.
I. Observing and data analysis systems, discovery and timing of 100 pulsars. *Monthly Notices
of the Royal Astronomical Society*, 328, 17–35.

Mannix, L., 2018. Chasing the dawn. *The Sunday Age*, 15 July, pp. 19–21.

Mills, B.Y., 2006. An engineer becomes an astronomer. *Annual Review of Astronomy and
Astrophysics* 44, 1–15.

Ng, C.-Y., Gaensler, B.M., Staveley-Smith, L., Manchester, R.N., Kesteven, M.J., Ball,
L., and Tzioumis, A.K., 2008. Fourier modeling of the radio torus surrounding SN
1987A. *Astrophysical Journal*, 684, 481–497.

Norris, R.P., and Kesteven, M.J., 2013. The life and times of the Parkes–Tidbinbilla Interferometer.
Journal of Astronomical History and Heritage, 16, 55–66.

Norris, R.P., and Norris, C., 2009. *Emu Dreaming: An Introduction to Australian Aboriginal
Astronomy*. Sydney, Emu Dreaming Publishing.

Orchiston, W., 2017. Cook, Green, Maskelyne and the 1769 transit of Venus: the legacy of the
Tahitian observations. *Journal of Astronomical History and Heritage*, 20, 35–68.

Orchiston, W., and Mathewson, D., 2009. Chris Christiansen and the Chris Cross. *Journal of
Astronomical History and Heritage*, 12, 11–32.

Orchiston, W., and Slee, B., 2017. The early development of Australian radio astronomy: the
role of the CSIRO Division of Radiophysics field stations. In Nakamura, T., and Orchiston,
W. (eds), *The Emergence of Astrophysics in Asia: Opening a New Window on the Universe*.
Cham (Switzerland), Springer. Pp. 497–578.

Peterson, B.A., Savage, A., Jauncey, D.L., and Wright, A.E., 1982. PKS 2000–330: A quasi-stellar
source with a redshift of 3.78. *Astrophysical Journal*, 260, L27–29.

Robertson, P., 1986. Grote Reber – Last of the lone mavericks. *Search*, 17, 118–121.

Robertson, P., 1992. *Beyond Southern Skies – Radio Astronomy and the Parkes Telescope*. Sydney,
Cambridge University Press.

Robertson, P., 2017. *Radio Astronomer – John Bolton and a New Window on the Universe*. Sydney,
NewSouth Publishing.

Robinson, B.J., Cooper, B.F.C., Gardiner, F.F., Wielebinski, R., and Landecker, T.L., 1968.
Measurements of the pulsed radio source CP 1919 between 85 and 2700 MHz. *Nature*, 218,
1143–1145.

Sim, H., 2013. The Compact Array turns 25. *Australian Sky & Telescope*, Aug/Sept, pp. 38–43.

Sullivan, W.T., 1988. Early years of Australian radio astronomy. In Home, R.W. (ed.) *Australian
Science in the Making*. Cambridge, Cambridge University Press. Pp. 308–44.

Sullivan, W.T., 2009. *Cosmic Noise – A History of Early Radio Astronomy*. Cambridge, Cambridge
University Press.

Turtle, A.J., Campbell-Wilson, D., Bunton, J.D., Jauncey, D.L., Kesteven, M.J., Manchester, R.N.,
Norris, R.P., Storey, M.C., and Reynolds, J.E., 1987. A prompt radio burst from supernova
1987A in the Large Magellanic Cloud. *Nature*, 327, 38–40.

Wyithe, S., 2019. President's report to the ASA Annual General Meeting for 2018–2019, University
of Queensland, 10 July 2019.

Zhang, S.-B., Dai, S., Hobbs, G., Staveley-Smith, L., Manchester, R.N., Russell, C.J., Zanardo, G.,
and Wu, X.-F., 2018. Search for a radio pulsar in the remnant of supernova 1987A. *Monthly
Notices of the Royal Astronomical Society*, 479, 1836–1841.

About the Authors

Professor Wayne Orchiston works at the National Astronomical Research Institute of Thailand (NARIT) and is an Adjunct Professor of Astronomy in the Centre of Astrophysics at the University of Southern Queensland in Toowoomba, Australia. He has wide-ranging research interests that include the history of radio astronomy, and he has written papers about Australian, French, Indian, Japanese, New Zealand and US radio astronomy and supervised one MAstr and four PhD theses on the history of radio astronomy. He has also published extensively on historic transits of Venus and solar eclipses; historic telescopes and observatories; the history of cometary astronomy; the history of meteoritics; and ethnoastronomy.

Wayne's recent books include *Eclipses, Transits, and Comets of the Nineteenth Century: How America's Perception of the Skies Changed* (Springer, 2015, co-authored by Stella Cottam); *Exploring the History of New Zealand Astronomy: Trials, Tribulations, Telescopes and Transits* (Springer, 2016); *John Tebbutt: Rebuilding and Strengthening the Foundations of Australian Astronomy* (Springer, 2017); *The Emergence of Astrophysics in Asia: Opening a New Window on the Universe* (Springer, 2017, co-edited by Tsuko Nakamura); *Exploring the History of Southeast Asian Astronomy: A Review of Current Projects and Future Prospects and Possibilities* (Springer, 2021, co-edited by Mayank Vahai); and a succession of conference proceedings.

While working at NARIT Wayne's research interests have focused mainly on Asia, and especially Indian, Indonesian, Japanese, Philippines, Thai and Vietnamese astronomy, and astronomical links between SE Asia and India and SE Asia and

W. Orchiston et al., *Golden Years of Australian Radio Astronomy*,
Historical & Cultural Astronomy, https://doi.org/10.1007/978-3-319-91843-3

Australia. Wayne has been a member of the IAU since 1985, and has been very active in commissions dealing with history of astronomy, radio astronomy, and education and development. In 2003 he founded the IAU Working Group on Historical Radio Astronomy, and currently he is the President of Commission C3 (History of Astronomy). In 1998 he co-founded the *Journal of Astronomical History and Heritage*, and is the current Managing Editor. He also co-edits Springer's Series on Historical and Cultural Astronomy, and is the Radio Astronomy Section Editor of Springer's Third Edition of the *Biographical Encyclopedia of Astronomers*. Along with one of his former US graduate students, Dr Stella Cottam, Wayne was the co-recipient of the American Astronomical Society's 2019 Donald E. Osterbrock Book Prize, and minor planet '48471 Orchiston' is named after him.

Dr Peter Robertson developed an interest in the history of science while working at the Niels Bohr Institute in Copenhagen. The Institute was at the centre of the revolutionary new quantum theory developed in the 1920s, as described in Peter's book *The Early Years – The Niels Bohr Institute 1921–30* (Akademisk Forlag, 1979).

Peter spent most of his career (1980–2001) as managing editor of the *Australian Journal of Physics*, published jointly by CSIRO and the Australian Academy of Science. The *AJP* published the great majority of the research papers produced by the Radiophysics radio astronomers in the 1940s and 1950s, as described in Chapters 3 and 4 of this book. Peter has also served as the production editor (1993–2001) of the research journal *Publications of the Astronomical Society of Australia* and as editor of *Australian Physics* magazine (2010–12) published by the Australian Institute of Physics.

Peter has published two books on the history of Australian astronomy: *Beyond Southern Skies – Radio Astronomy and the Parkes Telescope* (Cambridge University Press, 1992) and *Radio Astronomer – John Bolton and a New Window on the Universe* (NewSouth Publishing, 2017). After his retirement in 2009, Peter completed a PhD at the University of Southern Queensland (co-supervised by Wayne Orchiston) on John Bolton and his role in the discovery of the first discrete radio sources at the Dover Heights field station (see Section 4.1 of this book).

Peter is currently an associate editor of the *Journal of Astronomical History and Heritage* and is an honorary research fellow in the School of Physics at the University of Melbourne.

Woodruff T. Sullivan III is Professor Emeritus in the Department of Astronomy and Adjunct Professor in the Department of History of the University of Washington (UW), Seattle, USA. He retired from the UW faculty in 2013 after teaching and researching for 40 years in the fields of Astronomy, Astrobiology, and History of Science. He majored in Physics as an undergraduate at the Massachusetts Institute of Technology and obtained his Ph.D. (Astronomy) in 1971 at the University of Maryland. His dissertation, under the supervision of Australian radio astronomy pioneer Frank Kerr, was titled *Microwave Radiation of Water Vapor in Galactic Sources*.

Woody has written *Cosmic Noise: A History of Early Radio Astronomy* (2009) and edited two earlier, related books: *Classics in Radio Astronomy* (1982) and *The Early Years of Radio Astronomy: Reflections Fifty Years after Jansky's Discovery* (1984). He is a Fellow of the American Astronomical Society and in 2012 was awarded the LeRoy E. Doggett Prize for lifetime achievement in the history of astronomy by its Historical Astronomy Division. For the past decade he has been focused on the research and writing of a biography of William Herschel (1738–1822).

Woody is also the past director of the UW's graduate interdisciplinary astrobiology program. His research in astrobiology has centred on the search for extraterrestrial intelligence (SETI), for instance, as co-founder of the pioneering seti@home project (1999). He was the first to evaluate the detailed radio 'leakage' escaping into the Galaxy from transmitters around the Earth (1978), and also to assemble a map of 'Earth at Night' (1986). With John Baross, he co-edited the graduate textbook *Planets and Life: The Emerging Science of Astrobiology* (2007). Sundials are his passion. He has designed many sundials in the Seattle region, further afield in the USA (including one at the Jansky Very Large Array in New Mexico), on the Martian rovers *Spirit* and *Opportunity*, and tattooed on his wrist.

Bibliography on Early Australian Radio Astronomy

Bell Burnell, S.J., 1977. Little Green Men, white dwarfs, or pulsars? *Annals of the New York Academy of Science*, 302, 685–689.

Bolton, J.G., 1982. Radio astronomy at Dover Heights. *Proceedings of the Astronomical Society of Australia*, 4, 349–358.

Bowen, E.G., 1984. The origins of radio astronomy in Australia. Sullivan, W.T., 1984. Pp. 85–111.

Bowen, E.G., 1987. *Radar Days*. Bristol, Adam Hilger.

Bowen, E.G., 1988. From wartime radar to postwar radio astronomy in Australia. *Journal of Electrical and Electronic Engineering (Australia)*, 8, 1–11.

Bracewell, R.N., 1984. Early work on imaging theory in radio astronomy. Sullivan, W.T., 1984. Pp. 167–190.

Bracewell, R.N., 2005. Radio astronomy at Stanford. *Journal of Astronomical History and Heritage*, 8, 75–86.

Brooks, J., and Sinclair, M., 1994. Receivers, electronics and people – past and present. In Goddard, D.E., and Milne, D.K. (eds). *Parkes – Thirty Years of Radio Astronomy*. Melbourne, CSIRO Publishing. Pp. 47–56.

Campbell, D.B., 2019. Radio astronomy at Cornell University: the early years, 1946 to 1962. *Journal of Astronomical History and Heritage*, 22, 503–520.

Christiansen, W.N., 1984. The first decade of solar radio astronomy in Australia. In Sullivan, W.T., 1984. Pp. 113–131.

Christiansen, W.N., and Mills, B.Y., 1964. Biographical Memoirs [Joseph Lade Pawsey]. *Australian Physicist*, 1, 137–141.

Cohen, M.H., 1994. The Owens Valley Radio Observatory: The early years. *Engineering and Science*, 57(3), 9–23.

Collis, B., 2002. *Fields of Discovery: Australia's CSIRO*. Sydney, Allen & Unwin.

Davies, R.D., 2005. A history of the Potts Hill radio astronomy field station. *Journal of Astronomical History and Heritage*, 8, 87–96.

Davies, R.D., 2009. Recollections of two and a half years with 'Chris' Christiansen. *Journal of Astronomical History and Heritage*, 12, 4–10.

Débarbat, S., Lequeux, J., and Orchiston, W., 2007. Highlighting the history of French radio astronomy. 1: Nordmann's attempt to observe solar radio emission in 1901. *Journal of Astronomical History and Heritage*, 10, 3–10.

Dillett, T., 2000. *The Royal Australian Air Force on Collaroy Plateau in the Second World War*. Collaroy (Sydney), J.E. & D.M. Dellit.

Edge, D.O., and Mulkay, M.J., 1976. *Astronomy Transformed: The Emergence of Radio Astronomy in Britain*. New York, Wiley.

© The Author(s) 2021
W. Orchiston et al., *Golden Years of Australian Radio Astronomy*,
Historical & Cultural Astronomy, https://doi.org/10.1007/978-3-319-91843-3

Ekers, R., 2013. The history of the Square Kilometre Array (SKA) born global. Proceedings of the Conference on Resolving the Sky – Radio Interferometry: Past, Present and Future. Manchester, UK, April 2012.

Franklin, K.L., 1984. The discovery of Jupiter bursts. In Kellermann, K.I., and Sheets, B., 1984. Pp. 252–257.

Frater, R.H., and Ekers, R.D., 2012. John Paul Wild AC CBE FAA FTSE. 17 May 1923 – 10 May 2008. *Biographical Memoirs of Fellows of the Royal Society*, 58, 327–346.

Frater, R.H., and Goss, M.W., 2011. Wilbur Norman Christiansen 1912–2007. *Historical Records of Australian Science*, 21, 215–228.

Frater, R.H., Goss, W.M., and Wendt, H.W., 2013. Biographical Memoir: Bernard Yarton Mills 1920 – 2011. *Historical Records of Australian Science*, 24, 294–315.

Frater, R.H., Goss, W.M., and Wendt, H.W., 2017. *Four Pillars of Radio Astronomy: Mills, Christiansen, Wild, Bracewell*. Cham (Switzerland), Springer.

Freeman, J., 1991. *A Passion for Physics: The Story of a Woman Physicist*. Bristol, IOP Publishing.

George, M., Orchiston, W., Slee, B., and Wielebinski, R., 2015a. The history of early low frequency radio astronomy in Australia. 2: Tasmania. *Journal of Astronomical History and Heritage*, 18, 14–22.

George, M., Orchiston, W., Slee, B., and Wielebinski, R., 2015b. The history of early low frequency radio astronomy in Australia. 3: Ellis, Reber and the Cambridge field station near Hobart. *Journal of Astronomical History and Heritage*, 18, 177–189.

George, M., Orchiston, W., Wielebinski, R., and Slee, B., 2015c. The history of early low frequency radio astronomy in Australia. 5: Reber and the Kempton field station near Hobart. *Journal of Astronomical History and Heritage*, 18, 312–324.

George, M., Orchiston, W., and Wielebinski, R., 2017. The history of early low frequency radio astronomy in Australia. 8: Grote Reber and the 'Square Kilometre Array' near Bothwell, Tasmania, in the 1960s and 1970s. *Journal of Astronomical History and Heritage*, 20, 195–210.

Goss, W.M., 2013. *Making Waves, The Story of Ruby Payne-Scott, Australian Pioneer Radio Astronomer*. Heidelberg, Springer.

Goss, W.M., and McGee, R.X., 2009. *Under the Radar – The First Woman in Radio Astronomy: Ruby Payne-Scott*. Heidelberg, Springer.

Goss, W.M., Ekers, R.D., and Hooker, C., 2021. *From the Sun to the Cosmos, Joseph Lade Pawsey, Founder of Australian Radio Astronomy*. Cham (Switzerland), Springer.

Hanbury Brown, R., Minnett, H.C., and White, F.W.G., 1992. Edward George Bowen 1911–1991. *Historical Records of Australian Science*, 9, 151–66.

Harris, M., 2017. *Rocks, Radio and Radar: The Extraordinary Scientific, Social and Military Life of Elizabeth Alexander*. New Jersey, World Scientific.

Haynes, R., Haynes, R., Malin, D., and McGee, R., 1996. *Explorers of the Southern Sky. A History of Australian Astronomy*. Cambridge. Cambridge University Press.

Hey, J.S., 1973. *The Evolution of Radio Astronomy*. London, Elek Science.

Jauncey, D.L., 2012. SHEVE '82, Tasso, and the coming age of Southern Hemisphere VLBI. Proceedings of the Conference on From Antikythera to the Square Kilometre Array: Lessons from the Ancients. Kerastari, Greece.

Jauncey, D.L., Preston, R.A., and (23 others), 1983. SHEVE: The Southern Hemisphere VLBI experiment. *Bulletin of the American Astronomical Society*, 15, 648.

Kellermann, K.I., 1996. John Gatenby Bolton (1922–1993). *Publications of the Astronomical Society of the Pacific*, 108, 729–737.

Kellermann, K.I., 2005. Grote Reber: radio astronomy pioneer. In Orchiston, W., 2005a. Pp. 43–70.

Kellermann, K.I., and Sheets, B. (eds), 1984. *Serendipitous Discoveries in Radio Astronomy*. Green Bank (WV), National Radio Astronomy Observatory.

Kellermann, K.I., Bouton, E.N., and Brandt, S.S., 2020. *Open Skies: The National Radio Astronomy Observatory and Its Impact on US Radio Astronomy*. Cham (Switzerland), Springer.

Kellermann, K.I., Orchiston, W., and Slee, B., 2005. Gordon James Stanley and the early development of radio astronomy in Australia and the United States. *Publications of the Astronomical Society of Australia*, 22, 13–23.

Kerr, F.J., 1984. Early days in radio and radar astronomy in Australia. In Sullivan, W.T., 1984a. Pp. 133–145.

Lomb, N., 2015. How astronomers focussed the scope of their discussions: The formation of the Astronomical Society of Australia. *Historical Records of Australian Science*, 26, 36–57.

Lovell, A.C.B., 1964. Joseph Lade Pawsey, 1908–1962. *Biographical Memoirs of Fellows of the Royal Society*, 10, 228–243.

Lovell, B., 1968. *The Story of Jodrell Bank*. New York, Harper & Row.

McLean, D.J., and Labrum, N.R. (eds), 1985. *Solar Radiophysics – Studies of Emission from the Sun at Metre Wavelengths*. Cambridge, Cambridge University Press.

McNally, D., 1990. Obituary. C.W. Allen (1904–1987). *Quarterly Journal of the Royal Astronomical Society*, 31, 259–266.

Melrose, D.B., and Minnett, H.C., 1998. Jack Hobart Piddington 1910–1997. *Historical Records of Australian Science*, 12, 229–246.

Mills, B.Y., 1985. Obituary – Little, Alec. *Proceedings of the Astronomical Society of Australia*, 6, 113.

Mills, B.Y., 2006. An engineer becomes an astronomer. *Annual Review of Astronomy and Astrophysics*, 44, 1–15.

Milne, D.K., 1994. John Bolton and the rainmcakers. *Australian Journal of Physics*, 47, 549–553.

Milne, D.K., and Whiteoak, J.B., 2005. The impact of F.F. Gardner on our early research with the Parkes Radio Telescope. *Journal of Astronomical History and Heritage*, 8, 33–38.

Morton, D.C., 1985. The centre of our Galaxy: Is it a black hole? *Australian Physicist*, 22, 218–222.

Nakamura, T., and Orchiston, W. (eds), 2017. *The Emergence of Astrophysics in Asia: Opening a New Window on the Universe*. Cham (Switzerland), Springer.

Norris, R.P., and Kesteven, M.J., 2013. The life and times of the Parkes–Tidbinbilla Interferometer. *Journal of Astronomical History and Heritage*, 16, 55–66.

Norris, R.P., and Norris, C., 2009. *Emu Dreaming: An Introduction to Australian Aboriginal Astronomy*. Sydney, Emu Dreaming Publishing.

Orchiston, W., 1994. John Bolton, discrete sources, and the New Zealand field-trip of 1948. *Australian Journal of Physics*, 47, 541–547.

Orchiston, W., 2004a. From the solar corona to clusters of galaxies: the radio astronomy of Bruce Slee. *Publications of the Astronomical Society of Australia*, 21, 23–71.

Orchiston, W., 2004b. The 1948 solar eclipse and the genesis of radio astronomy in Victoria. *Journal of Astronomical History and Heritage*, 7, 118–121.

Orchiston, W., 2004c. The rise and fall of the Chris Cross: a pioneering Australian radio telescope. In Orchiston, W., et al. (eds). *Astronomical Archives and Instruments in the Asia–Pacific Region*. Seoul, Yonsei University. Pp. 157–162.

Orchiston, W. (ed.), 2005a. *The New Astronomy: Opening the Electromagnetic Window and Expanding our View of Planet Earth*. Dordrecht, Springer.

Orchiston, W., 2005b. Dr Elizabeth Alexander: first female radio astronomer. In Orchiston, W., 2005a. Pp. 71–92.

Orchiston, W., 2005c. Sixty years in radio astronomy: a tribute to Bruce Slee. *Journal of Astronomical History and Heritage*, 8, 3–10.

Orchiston, W., 2012. The Parkes 18-m antenna: a brief historical evaluation. *Journal of Astronomical History and Heritage*, 15, 96–99.

Orchiston, W., 2016a. *Exploring the History of New Zealand Astronomy: Trials, Tribulations, Telescopes and Transits*. Cham (Switzerland), Springer.

Orchiston, W., 2016b. Introduction. In Orchiston, W., 2016a. Pp. 1–29.

Orchiston, W., 2016c. Elizabeth Alexander and the mysterious 'Norfolk Island Effect'. In Orchiston, W., 2016a, Pp. 629–651.

Orchiston, W., 2016d. John Bolton, Gordon Stanley, Bruce Slee and the riddle of the 'Radio Stars'. In Orchiston, 2016a. Pp.653–671.

Orchiston, W., 2017. Cook, Green, Maskelyne and the 1769 transit of Venus: the legacy of the Tahitian observations. *Journal of Astronomical History and Heritage*, 20, 35–68.

Orchiston, W., and Mathewson, D., 2009. Chris Christiansen and the Chris Cross. *Journal of Astronomical History and Heritage*, 12, 11–32.

Orchiston, W., and Phakatkar, S., 2019. A tribute to Professor Govind Swarup, FRS: The father of Indian radio astronomy. *Journal of Astronomical History and Heritage*, 22, 3–44.

Orchiston, W., and Robertson, P., 2017. The origin and development of extragalactic radio astronomy: the role of the CSIRO's Division of Radiophysics Dover Heights Field Station in Sydney. *Journal of Astronomical History and Heritage*, 20, 289–312.

Orchiston, W., and Slee, B., 2002. Ingenuity and initiative in Australian radio astronomy: the Dover Heights 'hole-in-the-ground' antenna. *Journal of Astronomical History and Heritage*, 5, 21–34.

Orchiston, W., and Slee, B., 2006. Early Australian observations of historic supernova remnants at radio wavelengths. In Chen, K.-Y., et al. (eds). *Proceedings of the Fifth International Conference on Oriental Astronomy*. Chiang Mai (Thailand), University of Chiang Mai. Pp. 43–56.

Orchiston, W., and Slee, B., 2017. The early development of Australian radio astronomy: the role of the CSIRO Division of Radiophysics field stations. In Nakamura, T., and Orchiston, W., 2017. Pp. 497–578.

Orchiston, W., and Wendt, H., 2017. The contribution of the Georges Heights experimental radar antenna to Australian radio astronomy. *Journal of Astronomical History and Heritage*, 20, 313–340.

Orchiston, W., George, M., Slee, B., and Wielebinski, R., 2015a. The history of early low frequency radio astronomy in Australia. 1: The CSIRO Division of Radiophysics. *Journal of Astronomical History and Heritage*, 18, 3–13.

Orchiston, W., Lequeux, J., Steinberg, J.-L., and Delannoy, J., 2007. Highlighting the history of French radio astronomy. 3: The Würzburg antennas at Marcoussis, Meudon and Nançay. *Journal of Astronomical History and Heritage*, 10, 221–245.

Orchiston, W., Nakamura, T., and Strom, R. (eds), 2011. *Highlighting the History of Astronomy in the Asia–Pacific Region*. New York, Springer.

Orchiston, W., Sim, H., and Robertson, P., 2016. Dr Owen Bruce Slee, 10 August 1924 – 18 August 2016. *Australian Physics*, 53, 214.

Orchiston, W., Slee, B., and Burman, R., 2006. The genesis of solar radio astronomy in Australia. *Journal of Astronomical History and Heritage*, 9, 35–56.

Orchiston, W., Slee, B., George, M., and Wielebinski, R., 2015b. The history of early low frequency radio astronomy in Australia. 4: Kerr, Shain, Higgins and the Hornsby Valley field station near Sydney. *Journal of Astronomical History and Heritage*, 18, 285–311.

Pawsey, J.L., 1960. Obituary Notices. Charles Alexander Shain. *Quarterly Journal of the Royal Astronomical Society*, 1, 244–245.

Pawsey, J. L., and Bracewell, R.N., 1955. *Radio Astronomy*. Oxford, The Clarendon Press.

Piddington, J.H., and Oliphant, M.L., 1972. David Forbes Martyn. *Records of the Australian Academy of Science*, 2(2), 47–60.

Reber, G., 1984. Radio astronomy between Jansky and Reber. In Kellermann, K.I., and Sheets, B., 1984. Pp. 71–78.

Robertson, P., 1986. Grote Reber: Last of the lone mavericks. *Search*, 17, 118–121.

Robertson, P., 1992. *Beyond Southern Skies: Radio Astronomy and the Parkes Telescope*. Sydney, Cambridge University Press.

Robertson, P., 2000. Joseph Lade Pawsey (1908–1962): founder of Australian radio astronomy. In Ritchie, J. (ed.). *Australian Dictionary of Biography, Volume 15*. Melbourne, Melbourne University Press. Pp. 578–579.

Robertson, P., 2002. Stefan Friedrich Smerd (1916–1978): CSIRO solar physicist. In Ritchie, J. (ed.). *Australian Dictionary of Biography, Volume 16*. Melbourne, Melbourne University Press. P. 266.

Robertson, P., 2016. John Bolton and the Nature of Discrete Radio Sources. PhD thesis, University of Southern Queensland, Queensland.

Robertson, P., 2017. *Radio Astronomer – John Bolton and a New Window on the Universe*. Sydney, NewSouth Publishing.

Robertson, P., and Bland-Hawthorn, J., 2014. Centre of the Galaxy: Sixtieth anniversary of an Australian discovery. *Australian Physics*, 51, 194–199.

Robertson, P., Cozens, G., Orchiston, W., and Slee, B., 2010. Early Australian optical and radio observations of Centaurus A. *Publications of the Astronomical Society of Australia*, 27, 402–430.

Robertson, P., Orchiston, W., and Slee, B., 2014. John Bolton and the discovery of discrete radio sources. *Journal of Astronomical History and Heritage*, 17, 283–306.

Shouguan, W., 2017. The early development of Chinese radio astronomy: the role of W.N. Christiansen. In Nakamura, T., and Orchiston, W., 2017. Pp. 245–254.

Sim, H., 2013. The Compact Array turns 25. *Australian Sky & Telescope*, Aug/Sept, Pp. 38–43.

Slee, O.B., 1994. Some memories of the Dover Heights field station, 1946–1954. *Australian Journal of Physics*, 47, 517–534.

Slee, O.B., 2005. Early Australian measurements of angular structure in discrete radio sources. *Journal of Astronomical History and Heritage*, 8, 97–116.

Stanley, G.J., 1994. Recollections of John Bolton at Dover Heights and Caltech. *Australian Journal of Physics*, 47, 507–516.

Stewart, R.T., 2009. The Contribution of the CSIRO Division of Radiophysics Penrith and Dapto Field Stations to International Radio Astronomy. PhD thesis, James Cook University, Queensland.

Stewart, R., Orchiston, W., and Slee, B., 2011a. The contribution of the Division of Radiophysics Dapto field station to solar radio astronomy, 1952–1964. In Orchiston, W., et al., 2011. Pp. 481–526.

Stewart, R., Orchiston, W., and Slee, B., 2011b. The Sun has set on a brilliant mind: Paul Wild (1923–2008). In Orchiston, W., et al., 2011. Pp. 527–542.

Stewart, R., Wendt, H., Orchiston, W., and Slee, B., 2010. The Radiophysics field station at Penrith, New South Wales, and the world's first solar radio spectrograph. *Journal of Astronomical History and Heritage*, 13, 2–15.

Stewart, R., Wendt, H., Orchiston, W., and Slee, B., 2011. A retrospective view of Australian solar radio astronomy 1945 to 1960. In Orchiston, W. et al., 2011. Pp. 589–629.

Sullivan, W.T., 1982. *Classics in Radio Astronomy*. Cambridge, Cambridge University Press.

Sullivan, W.T. (ed.), 1984a. *The Early Years of Radio Astronomy. Reflections Fifty Years after Jansky's Discovery*. Cambridge, Cambridge University Press.

Sullivan, W.T., 1984b. Karl Jansky and the discovery of extraterrestrial radio waves. In Sullivan, W.T., 1984a. Pp. 3–42.

Sullivan, W.T., 1984c. Karl Jansky and the beginnings of radio astronomy. In Kellermann, K.I., Sheets, B., 1984. Pp. 39–56.

Sullivan, W.T., 1988a. Early years of Australian radio astronomy. In Home, R.W. (ed.). *Australian Science in the Making*. Cambridge, Cambridge University Press. Pp. 308–44. Reprinted with minor changes in (2005) *Journal of Astronomical History and Heritage*, 8, 11–32. Also reprinted in (2017) Nakamura, T., and Orchiston, W., Pp. 453–496.

Sullivan, W.T., 1988b. Frank Kerr and radio waves: from wartime radar to interstellar atoms. In Blitz, L., and Lockman, F.J. (eds). *The Outer Galaxy*. New York, Springer. Pp. 268–287.

Sullivan, W.T., 1990. The entry of radio astronomy into cosmology: radio stars and Martin Ryle's 2C survey. In Bertotti, R., et al. (eds). *Modern Cosmology in Retrospect*. Cambridge, Cambridge University Press. Pp. 309–330.

Sullivan, W.T., 2009. *Cosmic Noise: A History of Early Radio Astronomy*. Cambridge, Cambridge University Press.

Swarup, G., 2006. From Potts Hill (Australia) to Pune (India): The journey of a radio astronomer. *Journal of Astronomical History and Heritage*, 9, 21–34.

Swarup, G., 2008. Reminiscences regarding Professor W.N. Christiansen. *Journal of Astronomical History and Heritage*, 11, 194–202.

Thomas, B.M., and Robinson, B.J., 2005. Harry Clive Minnett, 1917–2003. *Historical Records of Australian Science*, 16, 199–220.

Thompson, A.R., and Frater, R., 2010. Ronald N. Bracewell: an appreciation. *Journal of Astronomical History and Heritage,* 13, 172–178.

Van Woerden, H., and Strom, R., 2006. The beginnings of radio astronomy in The Netherlands. *Journal of Astronomical History and Heritage*, 9, 3–20.

Wendt, H., 2008. The Contribution of the Division of Radiophysics Potts Hill and Murraybank Field Stations to International Radio Astronomy. PhD thesis, James Cook University, Queensland.

Wendt, H., 2011. Paul Wild and his investigation of the H-line. In Orchiston, W., et al., 2011. Pp. 543–546.

Wendt, H., and Orchiston, W., 2018. The contribution of the AN/TPS-3 radar antenna to Australian radio astronomy. *Journal of Astronomical History and Heritage*, 21, 65–80.

Wendt, H., and Orchiston, W., 2019. The short-lived CSIRO Division of Radiophysics field station at Bankstown Aerodrome in Sydney. *Journal of Astronomical History and Heritage*, 22, 266–272.

Wendt, H., Orchiston, W., and Slee, B., 2008a. The Australian solar eclipse expeditions of 1947 and 1949. *Journal of Astronomical History and Heritage*, 11, 71–78.

Wendt, H., Orchiston, W., and Slee, B., 2008b. W.N. Christiansen and the development of the solar grating array. *Journal of Astronomical History and Heritage*, 11, 173–184.

Wendt, H., Orchiston, W., and Slee, B., 2008c. W.N. Christiansen and the initial Australian investigation of the 21-cm hydrogen line. *Journal of Astronomical History and Heritage*, 11, 185–193.

Wendt, H., Orchiston, W., and Slee, B., 2011a. The contribution of the Division of Radiophysics Potts Hill field station to international radio astronomy. In Orchiston, W., et al., 2011. Pp. 379–431.

Wendt, H., Orchiston, W., and Slee, B., 2011b. The contribution of the Division of Radiophysics Murraybank field station to international radio astronomy. In Orchiston, W., et al., 2011. Pp. 433–479.

Wendt, H., Orchiston, W., and Slee, B., 2011c. An overview of W.N. Christiansen's contribution to Australian radio astronomy, 1948–1960. In Orchiston, W., et al., 2011. Pp. 547–587.

Westerhout, G., 2000. Obituary: Frank John Kerr, 1918–2000. *Bulletin of the American Astronomical Society*, 32, 1674–1676.

Westfold, K.C., 1994. John Bolton – some early memories. *Australian Journal of Physics*, 47, 535–539.

Whiteoak, J.B., and Sim, H.L., 2006. Brian John Robinson 1930–2004. *Historical Records of Australian Science*, 17, 263–281.

Wild, J.P., 1980. The Sun of Stefan Smerd. In Kundu, M.R., and Gergely, T.E. (eds). *Radio Physics of the Sun*. Dordrecht, Reidel. Pp. 5–21.

Wild, J.P., and Radhakrishnan, V., 1995. John Gatenby Bolton, 5 June 1922 – 6 July 1993. *Biographical Memoirs of Fellows of the Royal Society*, 41, 72–86.

Index

Page numbers in *italics* refer to illustrations.

A

Ables, J., *225*
Aboriginal astronomy, 227
Aldrin, B., 235
Alexander, E., 7, 8, 102–104, *104*, 105
Alfvén, H., 173
Allen, C.W. ('Cla'), 16, 30, 108–110, *109*, 115
Andromeda Nebula (M31), 52, 173
Anglo-Australian Observatory (AAO), 239
Anglo-Australian Telescope (AAT), 9, 138, 235, 237
Aperture synthesis, viii, 223
Apollo program, 9, 138, 234, 235
Appleton, E., 8, 24, *24*, *26*, 32, 136
Armstrong, N., 235
Astronomical Society of Australia (ASA), 128, 176, 180, 238
Astrophysical Journal, 6, 16, 101, 135, 190
Attwood, C., *28*, *103*
Australia Telescope (AT), xv, 33, 176, 224, *226*, 227–231, *230*, 245
Australia Telescope Compact Array (ATCA), xi, 159, 191, *215*, *228*, *230*, *233*, *236*, 237, 247
Australia Telescope National Facility (ATNF), ix, xiv–xvi, *14*, 138, 159, 191, *226*, *229*, 229, *233*, *236*, 237, 243
See also Radiophysics Laboratory
Australian Academy of Science, 13, 45, 46, 66, 75, 81, 107, 220, *223*, 252
Australian Journal of Physics, ix, 32, 120, 167, 180, 192, 252
Australian Journal of Scientific Research, 32, 111, 115, 117, 118, 151, *152*, 161, 178, 183, 189
Australian National University, 13, *29*, 29, 229, 237, 238
Australian Square Kilometre Array Pathfinder (ASKAP), xi, *236*, *240*, 241, *241*, 242, *242*
Australian Synthesis Telescope (AST), 223–225, *225*, *226*, 227, 228

B

Baade, W., 16, 161, *162*, 173, 180
Badgery's Creek field station, *14*, *41*, 62, 80, 81, *82*, *83*, 84, 161, 163, *164*, 166, 173, 231
Bankstown Aerodrome field station, *14*, *41*, 62
Bell Burnell, J., 220
Bell Telephone Laboratories, 1, 3
Beringer, R.E., 135, 136
Big Bang theory, 168, *171*
See also Steady State theory
Binskin, L., *103*
Blaauw, A., 193
Bogle, G. ('Gib'), *80*
Bok, B.J., 13
Bolton, J.G., vii–x, xv, 13, 16, *24*, *25*, 27, *28*, 46, *46*, 81, *156*, 220, *225*, 229, 252
at Caltech, *19*, 121, 160, 171, 207
at Dover Heights, 18, 44, 47, 48, *48*, 50, 51, 55, 110, 111, 149, *150*, 151, *151*, *152*, 153, 155, 157, 160, 161, 164–166, 180, *181*, 183, 231
at Parkes Observatory, 217, 218, 225

© The Author(s) 2021
W. Orchiston et al., *Golden Years of Australian Radio Astronomy*,
Historical & Cultural Astronomy, https://doi.org/10.1007/978-3-319-91843-3